Secondary Research Methods in the Built Environment

The use of secondary data for research can offer benefits, particularly when limited resources are available for conducting research using primary methods. Researchers and students at both undergraduate and postgraduate levels, including their academic instructors, are increasingly recognising the immense opportunities in applying secondary research methods in built environment research. Advances in technology have also led to vast amounts of existing datasets that can be utilized for secondary research. This textbook provides a systematic guide on how to apply secondary research methods in the built environment, including their various underpinning methodologies. It provides guidance on the secondary research process, benefits, and drawbacks of applying secondary research methods, how to source for secondary data, ethical considerations, and the various secondary research methods that can be applied in built environment research. The book incorporates chapters dealing with qualitative secondary analysis, systematic literature reviews, legal research, bibliometric and scientometric analysis, literature-based discovery, and meta-analysis.

Secondary Research Methods in the Built Environment is an ideal research book for undergraduate and postgraduate students in construction management, construction project management, quantity surveying, construction law and dispute resolution, real estate and property management, building services engineering, architecture, and civil engineering.

Emmanuel Manu is an Associate Professor in Quantity Surveying and Project Management at the School of Architecture, Design and the Built Environment, Nottingham Trent University. He has supervised and examined several doctoral research projects in construction project management and engages in research consultancy, both in the UK and internationally. His research interests span the areas of construction supply chain management, circular economy in the built environment, sustainable procurement and social value, and smart and digital processes for performance improvement in the built environment. He has collaborated with industry partners to deliver research and consultancy projects totalling over £9 million, with two notable projects being on implementing Cradle to Cradle® principles (sustainability paradigm) in the built environment and the other on supply chain performance improvement. He is a member of the Association for Project Management and a Fellow of the UK Higher Education Academy.

Julius Akotia is a Senior Lecturer in Construction Project Management at the school of Architecture, Computing and Engineering (ACE), University of East London (UEL). He is also the link coordinator for the ACE academic 'international partnership' BSc Construction Management programmes with AMC metropolitan college in Greece, Kazan State University of Architecture and Engineering (KSUAE) in Russia, and Ain Shams University (ASU) in Egypt. He has supervised several undergraduate, postgraduate, and doctorial students and played external examiner role for national and international institutions. He is a chartered member of Chartered Institute of Building (MCIOB) and a fellow of the Higher Education Academy (FHEA). His research interests lie in the area of sustainability, with a focus on the social and economic sustainability aspects of sustainable regeneration. He has published over 18 peer-reviewed research papers in local and international conferences and academic journals.

Secondary Research Methods in the Built Environment

Edited by
Emmanuel Manu and Julius Akotia

Routledge
Taylor & Francis Group

LONDON AND NEW YORK

First published 2021
by Routledge
2 Park Square, Milton Park, Abingdon, Oxon OX14 4RN

and by Routledge
52 Vanderbilt Avenue, New York, NY 10017

Routledge is an imprint of the Taylor & Francis Group, an informa business

British Library Cataloguing-in-Publication Data
A catalogue record for this book is available from the British Library

Library of Congress Cataloging-in-Publication Data
Names: Manu, Emmanuel, editor. | Akotia, Julius, editor.
Title: Secondary research methods in the built environment /
edited by Emmanuel Manu and Julius Akotia.
Description: Abingdon, Oxon; New York, NY: Routledge/Taylor & Francis Group, 2021. | Includes bibliographical references and index.
Identifiers: LCCN 2020042567 (print) | LCCN 2020042568 (ebook) |
ISBN 9780367429881 (hardback) | ISBN 9780367429874 (paperback) |
ISBN 9781003000532 (ebook)
Subjects: LCSH: Building—Research. | Technical literature.
Classification: LCC TH213.5 .S43 2021 (print) |
LCC TH213.5 (ebook) | DDC 624.072—dc23
LC record available at https://lccn.loc.gov/2020042567
LC ebook record available at https://lccn.loc.gov/2020042568

ISBN: 978-0-367-42988-1 (hbk)
ISBN: 978-0-367-42987-4 (pbk)
ISBN: 978-1-003-00053-2 (ebk)

Typeset in Goudy
by codeMantra

Contents

Figures

Tables

Boxes

Contributors

Julius Akotia is a Senior Lecturer in Construction Project Management at the school of Architecture, Computing and Engineering (ACE), University of East London (UEL). He is also the link coordinator for the ACE academic 'international partnership' BSc Construction Management programmes with AMC metropolitan college in Greece, Kazan State University of Architecture and Engineering (KSUAE) in Russia, and Ain Shams University (ASU) in Egypt. He has supervised several undergraduate, postgraduate, and doctorial students and played external examiner role for national and international institutions. He is a chartered member of Chartered Institute of Building (MCIOB) and a fellow of the Higher Education Academy (FHEA). His research interests lie in the area of sustainability, with a focus on the social and economic sustainability aspects of sustainable regeneration. He has published over 18 peer-reviewed research papers in local and international conferences and academic journals..

Fernando Almeida is a computer science engineer. He is a full Professor at ISP-GAYA and researcher at University of Porto and INESC TEC. He performs research in the field of innovation, entrepreneurship, and software engineering by applying qualitative, quantitative, and mixed methods approaches.

Bankole Osita Awuzie is an Associate Professor at the Centre for Sustainable Smart Cities 4.0, Central University of Technology (CUT) Free State, South Africa. He holds a BSc degree in Real Estate Management from Imo State University, Owerri, Nigeria; an MSc degree in Construction Project Management from the Robert Gordon University, Aberdeen, Scotland; and a PhD in the Built Environment from the University of Salford, both in the United Kingdom. His research interests span the areas of smart, sustainable and circular built environments with several publications within this domain. He has successfully supervised several postgraduate students to completion at masters and doctoral degree levels, most of which have focused on studies in the aforementioned areas of research interests.

Nick Blismas is Professor of construction management in the School of Property, Construction and Project Management at RMIT University. He was an active member and past Director of the Centre for Construction Workplace

Health and Safety Research at RMIT. He also served on RMIT's Design and Social Context College Research Committee for several years. He has an extensive research record covering areas such as construction OHS and offsite manufacturing. He has completed several competitive research grant projects as well as numerous applied research projects funded by industry and other organisations.

Albert Ping-Chuen Chan is currently the Head of Department of Building and Real Estate and Associate Director of Research Institute for Sustainable Urban Development at The Hong Kong Polytechnic University. An Able Professor in Construction Health and Safety and a Chair Professor of Construction Engineering and Management, his research and teaching interests/expertise include but not limited to project management and project success, construction procurement and relational contracting, construction management and economics, construction health and safety, and construction industry development. He has published numerous articles including those using science mapping.

Daniel Wai-Ming Chan is an Associate Professor in Construction Project Management at the Department of Building and Real Estate, The Hong Kong Polytechnic University, Hong Kong. He is a Project Manager, Chartered Building Engineer, and Registered Construction Manager by profession. Also, he was the Honorary Secretary and Member of the Executive Council of the Hong Kong Institute of Project Management (HKIPM) between July 2013 and June 2017. He has published over 240 research papers on the broad theme of project management in leading construction management journals and international conference proceedings.

Vijayan Chelliah is a current PhD candidate in Department of Civil Engineering, Indian Institute of Technology Guwahati, India. His research interests include implementation of BIM and assessment of PPP projects.

Amos Darko is currently a Postdoctoral Fellow in the Department of Building and Real Estate at The Hong Kong Polytechnic University. He holds a PhD in Construction Management and Engineering from The Hong Kong Polytechnic University, and a BSc (First-class Honours) in Construction Technology and Management from Kwame Nkrumah University of Science and Technology. A passionate researcher and educator, his research and teaching interests/expertise include but not limited to sustainability in the built environment, green or sustainable building, modular and offsite construction, building information modelling, artificial intelligence, and research methods. He has published numerous articles including those using science mapping.

Ganesh Devkar is Associate Professor in Faculty of Technology, CEPT University, Ahmedabad, India. His research focuses on identifying the competencies necessary for effective implementing PPPs in infrastructure service delivery and developing a framework that allows urban bodies to assess their capability to successfully deliver projects through PPPs.

Heyecan Giritli, PhD is a retired Professor at the Faculty of Architecture at Istanbul Technical University, Turkey. She holds a BSc in Architecture, MSc in Construction Management, and PhD in Building Sciences. Her early research interests were in the areas of building economics. Current research interests include international construction, risk management, sustainability, and management of people in construction – especially cultural issues as drivers of behaviour. She wrote several papers on these subjects.

Abid Hasan is a Lecturer in the School of Architecture and Built Environment at Deakin University, Australia. In this role, he supervises both undergraduate and postgraduate students who often use secondary data in their research. His research interest includes human-technology interaction, health and safety of construction workers, project success factors, and productivity in construction projects.

Nimesha S. Jayasena is a full-time PhD research student at the Department of Building and Real Estate, The Hong Kong Polytechnic University, Hong Kong, and working on developing Public-Private Partnership models for smart infrastructure development projects. She obtained her MSc Degree from the Department of Building Economics, University of Moratuwa, Sri Lanka, in 2019. Her research interests include Sustainable Development, Urban Planning, Smart Cities, Public-Private Partnerships, Risk Management, Green Procurement, and Performance Management.

Ruoyu Jin currently works as an Associate Professor in Construction Management at London South Bank University, UK. Ruoyu's main research interests focus on Building Information Modelling (BIM) pedagogy, construction safety education, construction waste management especially recycled aggregate concrete, and off-site manufacturing for construction.

Joseph Kangwa, PhD is an Associate Professor of Construction Management at London South Bank University. His research interests are in Sustainable construction, Low carbon technology, empowerment of construction SMEs for growth and capacity, Low carbon refurbishment and the conservation of heritage buildings, Health and Safety Management, and enforcement and its impact on built environment process, including the use of recycled waste as building materials and its complementarity with sustainable design.

Nathan Kibwami, PhD is a Lecturer in the Department of Construction Economics and Management, Makerere University, Uganda. He has taught and supervised undergraduate and postgraduate students in construction project management since 2007. He has keen and enduring interest in methods of research in the subject area of the built environment.

Boeing Laishram is Associate Professor in Department of Civil Engineering, Indian Institute of Technology Guwahati, India. He worked in an investment bank engaged in appraisal and structuring of project finance deals for infrastructure projects in India. His research interests include sustainability assessment of PPP, relational contracting, BIM in PPP projects, and MCDM in PPP project selection.

Helen Lingard is an RMIT Distinguished Professor and Director of a research team focused on construction work health and safety. Helen has undertaken extensive applied research in the areas of workplace safety, workers' health and well-being, and work-family interaction in the construction industry. Her work has been funded by many private and public sector construction organisations. Recent projects include an analysis and evaluation of client initiatives in driving work health and safety improvements in the planning, design, and construction of major transport infrastructure construction projects and an analysis of the cultural, organisational, and job design factors that impact on construction workers' physical and mental health.

Samantha Low-Choy is an applied statistician at Griffith University, Queensland, Australia, with a career-long interest in advising and collaborating with researchers to tailor statistical methodology to their problems; developing novel methodology where needed. Increasingly she works at the Qual/Quant interface, including eliciting expert knowledge, informative priors in Bayesian statistical models, multi-/mixed methods, conceptual modelling, graphical modelling.

Abdul-Majeed Mahamadu, PhD is a Senior Lecturer in Construction Project Management in the Department of Architecture and the Built Environment, University of the West of England, Bristol (UWE). He is a chartered member of the Institute of Building, UK, and has over ten years of industry, research, consultancy, and teaching experience in both construction and infrastructure. He is actively involved in the delivery of research projects mainly in the areas of Building Information Modelling (BIM), digitisation, and occupational health and safety in construction.

Joseph Mante, PhD, LLM, BL, LLB is a Lecturer in law and the Course leader of the LLM/MSc Construction Law and Arbitration (CLARB) programme at the Robert Gordon University with expertise in complex contracts, construction law, commercial law, and dispute resolution. He is a Fellow of the UK Higher Education Academy.

Emmanuel Manu is an Associate Professor in Project Management and Quantity Surveying at the School of Architecture, Design and Built Environment, Nottingham Trent University. He has over ten years of research, consultancy, and teaching experience in the construction sector. His teaching responsibilities cover the initiation of projects, project procurement, contract administration, and contract practice. His research interests span the areas of construction supply chain management, circular economy in the built environment, sustainable procurement and social value, and smart and digital processes for performance improvement in the built environment. He has collaborated with industry partners to deliver research and consultancy project totalling over £9 million, with two notable projects being on implementing Cradle to Cradle® principles (sustainability paradigm) in the built environment and the other on supply chain performance improvement. He is a member of the Association for Project Management and a Fellow of the UK Higher Education Academy.

Timothy O. Olawumi is a Postdoctoral Fellow with the Chinese National Engineering Research Centre for Steel Construction (CNERC) at the Department of Civil and Environmental Engineering, The Hong Kong Polytechnic University, Hong Kong. He completed his PhD at the Department of Building and Real Estate of the same university. He has over 34 publications in high-impact journals and conference proceedings on research topics such as BIM, green buildings, sustainability assessment, construction project management, modular (offsite) construction, blockchain technology, construction informatics, and construction economics. He holds full membership of the Hong Kong Green Building Council Limited (MHKGBC), buildingSMART International, and a Member of BIMAfrica®. He also has an Associate Membership of the Hong Kong Institute of Project Management (AHKIPM) and the Royal Institute of Chartered Surveyors. (RICS).

Olalekan Shamsideen Oshodi is a Lecturer in Construction Project Management at the School of Engineering and the Built Environment, Anglia Ruskin University, United Kingdom. Olalekan is a Member of the Chartered Institute of Building (MCIOB) and a Fellow of the UK Higher Education Academy (FHEA). He has received grants in excess of £200,000 towards his research. He has previously worked in tertiary institutions in Nigeria, Hong Kong, South Africa and the United Kingdom and has supervised and mentored several undergraduate and postgraduate students. His research interests include construction productivity and performance improvements, off-site construction, sustainability, enhancement of labour utilisation and application of data mining techniques to problems in the built environment. He has authored over 43 peer-reviewed publications in book chapters, conferences and academic journals.

Payam Pirzadeh, PhD is a Lecturer at RMIT University. He has participated in a number of research projects funded by industry and government, including safety in design, a supply-chain approach to health and safety, the influence of clients in driving health and safety improvement in construction projects, the impact of supervisors' and site managers' behaviours on health and safety, and the use of commercial frameworks to drive exceptional health and safety. He was the recipient of the RMIT Prize for Excellence in Research by a Higher Degree student in 2019 for his doctoral research in relation to 'safety in design' in construction.

Suresh Renukappa currently serves as a Senior Lecturer in the Faculty of Science and Engineering at the University of Wolverhampton. He holds a PhD in managing change and knowledge associated with sustainability initiatives for improved competitiveness. He has over 20 years of research, consultancy, project management, and teaching experience in a wide range of business and management areas across industrial sectors in both developed and emerging economies. His research interests cover, but not limited to, sustainability strategies for competitive advantage; carbon reduction strategies; corporate social responsibility; smart cities development; leading change towards sustainability; knowledge management; public private partnerships; cloud computing; infrastructure asset management; and sustainable infrastructure investment and development. He has successfully executed more than 30 large projects and authored over 100 papers which have been published in journals, book chapters, and conference proceedings.

Judy Rose is a social scientist and mixed methodologist in concurrent, sequential, or integrated conduct of qualitative and quantitative research methods. Judy has combined a quantitative systematic review of literature with qualitative discourse analysis and a phenomenological interpretivist approach. Judy teaches and writes on mixed methods in higher education.

Abdullahi B. Saka is a full-time PhD research student at the Department of Building and Real Estate, The Hong Kong Polytechnic University, Hong Kong; and working on Building Information Modelling Adoption and Implementation in Developing Countries. He is an Associate Member of the Hong Kong Institute of Project Management (AHKIPM). Also, he is currently the Lead for Research and Development Committee of BIM Africa, a Pan-African community of construction professionals and researchers with the responsibility of digitizing the African Built Environment.

Saad Sarhan, PhD is a programme leader and Senior Lecturer in construction management at University of Lincoln. Saad has strong passion for research and development. He is actively researching on latest trends in lean construction, construction project procurement, the concept of waste in construction, occupational stress in construction, and work related to improving supply-chain efficiency. Saad has recently won the 2019/2020 Building and Civil Engineering Charitable Trust's Occupational Health Research Award, an annual grant of £25,000 for research benefiting the UK construction industry.

Redouane Sarrakh is a PhD researcher at the Faculty of Science and Engineering, the University of Wolverhampton, researching the impact of sustainability strategies on the Qatar energy sector competitiveness. He received a BEng (Hons) in Process Engineering from the EMI Engineering School in Morocco and MSc in Oil and Gas Management from the University of Wolverhampton. His research interests include sustainability of the oil and gas industry, competitiveness, leading change towards sustainability, input-output modelling, interpretive structural modelling, energy subsidy reform, mobile applications, and big data analytics in energy sector.

Victoria Sherif, PhD is an Assistant Professor in the Department of Counseling, Educational Leadership, Education and School Psychology at Wichita State University, USA. Her research centres around issues related to qualitative secondary analysis and its application in social and education sciences. She is also interested in examination of leadership of students and educational administrators in early childhood, K-12, and school district settings.

Subashini Suresh, PhD has over 19 years of experience in research, teaching, and practice in the area of Project Management and has worked in the area of Architecture, Engineering and Construction (AEC) sector in UK, USA,

UAE, Nigeria, Ghana, Italy, Netherlands, and India. Currently, she is a Reader of Construction Project Management at the School of Architecture and Built Environment, University of Wolverhampton. She holds a PhD in knowledge management. She received Rewarding Excellence Award for Innovation in Teaching and also for Blended Learning Tutor. She has published over 150 academic publications, which include 27 journal papers, 95 conference papers, four articles, eight book chapters, 15 reports, and three books. Her key areas of interest are as follows: construction project management, knowledge management, building information modelling, health and safety, sustainability/ green construction, emerging technologies, quality management, leadership in change management initiatives, organisational competitiveness, business process improvement, lean construction, risk management, and Six Sigma leadership.

Ecem Tezel, MSc is a research assistant at Faculty of Architecture of Istanbul Technical University, Turkey. She holds a Bachelor's degree in architecture and MSc degree in Project and Construction Management. She has been studying her PhD in Project and Construction Management Program. During her master study, she focused on pro-environmental behaviours (PEBs) of occupants in sustainable and non-sustainable office buildings with the aim of investigating the effect of sustainable building certificates on users' actions in workplaces. Her current research interests include operation and maintenance phase, facilities management, retrofit/refurbishment projects, and BIM implementation.

Nicola Thounaojam is a current PhD candidate in Department of Civil Engineering, Indian Institute of Technology Guwahati, India. Her research interests include sustainability assessment in PPP, megaprojects, and institutional theory.

Apollo Tutesigensi, PhD is an Associate Professor in Infrastructure Project Management in the School of Civil Engineering, University of Leeds, UK. He has supervised numerous undergraduate, postgraduate, and research students since 1993. He has also examined numerous PhD candidates in UK and overseas on various topics in construction project management.

Liyuan Wang is a Lecturer in the Department of Civil Engineering at Fuzhou University, China. Specialising in Building Information Modelling education and pedagogical research in digital construction, Liyuan has published multiple academic papers in BIM education and construction engineering. Liyuan has also provided consulting service to multiple building and infrastructure construction projects by developing the digital communication platform incorporating BIM.

Preface

The inspiration for this book came about through a conversation between the two editors who recognised the challenges undergraduate and postgraduate students studying on built environment courses experience when completing their dissertations. Due to limited time and resources, students have often struggled to adequately gather and analyse data first-hand using primary research methods. However, we now live in a world full of data, with significant opportunities to apply secondary research methods for research in the built environment. The application of secondary research methods can offer an alternative to primary research methods as well as compliment it. Despite the possibilities of applying secondary research methods in built environment research, especially when there are time and resource constraints, conversations with students and academic colleagues alike have revealed that this is an area where clear guidance is needed. The secondary research domain can be a minefield for experienced researchers who are unfamiliar with secondary research methods, let alone the novice researcher. This is compounded by the fact that existing research methods books dedicated to the built environment disciplines have focused mainly on primary research methods. At best, some of these existing books only make mention of secondary research methods, without delving into any detail. Some of the questions we have often been confronted with during our engagements with students include whether the review of literature amounts to conducting secondary research, where data can be sourced for secondary research, which secondary research methods can be used to answer their research questions, what to include in the research methodology chapter of the dissertation if secondary research methods are adopted, what methods to use in analysing the secondary data, and how to report findings from the secondary analysis.

Our intention is that some, if not most, of these questions will be answered across the various chapters in the book. Some of the areas that have been covered in the book include the design of secondary research, sourcing of data for secondary research, ethical requirements to observe when conducting secondary research, qualitative secondary analysis, systematic literature reviews, legal research, bibliographic and scientometric analysis, literature-based discovery, meta-analysis, and the analysis of other pre-existing quantitative and qualitative datasets. In addition to addressing these areas, case examples have also been used to showcase some of the secondary methods. These case examples are aimed at guiding students and researchers who are interested in applying similar methods in their research.

It is our hope that this book will serve as a valuable resource for undergraduate and postgraduate students undertaking research across various built environment courses such as construction management, construction project management, quantity surveying, building services, construction law and dispute resolution, real estate and property management, and civil engineering.

Emmanuel Manu and Julius Akotia

1 Introduction to secondary research methods in the built environment

Emmanuel Manu and Julius Akotia

Introduction

Secondary research methods involve the analysis of data that already exists or has already been created. This is in contrast to primary research, which is based on principles of the scientific method (Driscoll, 2011) where researchers learn more about the world by collecting measurable data first-hand. In recent years, the use of secondary research methods has grown exponentially across various disciplines, including in the built environment. This growth has been attributed to technological advances (Johnston, 2014) and the vast amounts of secondary data that is now available and easily accessible for research as a result. Researchers and students at both undergraduate and postgraduate levels across various built environment disciplines, including their academic instructors, are increasingly recognising and exploiting the immense opportunities in conducting secondary research. However, for the inexperienced secondary researcher, some of the issues with which they are confronted include clarifications to questions such as: what exactly is secondary research? What types of secondary research methods are available, and how can these methods be applied robustly in built environment research? This is hardly surprising as issues relating to secondary research can be confusing for experienced researchers who are unfamiliar with secondary research methods, let alone the novice researcher. This situation has been complicated by the sometimes differing views and varying terminologies that are used to describe different secondary research methods. Terminologies, such as secondary data analysis, qualitative secondary analysis (QSA), qualitative secondary research, meta-analysis, and meta-synthesis, abound in the literature without clarity on how all these fit within the domain of secondary research methods.

Therefore, in this introductory chapter, the aim is to create a context for the rest of this book by evaluating what constitutes secondary research, the secondary research process, secondary research designs and the benefits and drawbacks of applying secondary research methods in built environment research. To conclude, an overview of the various chapters that are included in this book has been presented.

What is secondary research?

Secondary research involves the use of data that already exists rather than what would be obtained from first-hand sources, using primary methods such as questionnaires, interviews, focus groups, observations, and the like. Johnston (2014, p. 619) who used the term "secondary data analysis" defined this as "the analysis of data that [were] collected by someone else for another primary purpose".

There are subtle differences in perspectives as to what constitutes secondary research. In some discussions the re-analysis of data from previous primary research is emphasised, suggesting that all forms of secondary research will utilise data that have emerged from a primary research study, which have been analysed and published in academic literature, but are then being re-analysed. However, in the definition offered by Johnston (2014), the term "analysis" is used rather than "re-analysis", an indication perhaps that the data for secondary analysis might be in its raw state but would still have come into existence for a different purpose. Whilst some secondary methods are focused on the re-analysis of published findings from primary research (e.g. systematic literature reviews (SLRs) and meta-analysis), confining secondary research to only the use of such methods is a narrow perspective. This is because there are other established secondary research methods such as the QSA that involve the re-use of archived, original (pre-existing) qualitative data (e.g. interview transcripts) to answer new research questions. The raw qualitative data would have been archived in their original, pre-analysed, and unpublished form and, hence, their re-use as secondary data is a case of re-purposing the original data. With this practice, the existing unanalysed and archived, raw data are re-used as opposed to re-analysing the published results from a previous analysis of the same data.

Johnston (2014) also emphasised in his definition that the existing data should have been collected by someone else rather than the secondary researcher. This relationship between the secondary researcher and how the existing data came about has also featured in other discussions on secondary research. Church (2001), for instance, emphasised that in secondary data analysis, individuals that were not involved in data collection are responsible for analysing the data, unlike in primary research where individuals that collect the data are also responsible for analysing it. This suggests that secondary research will always involve the use of data that have been collected by someone else, which is mostly the case. However, it is important to clarify also that secondary research can involve the re-use of existing raw data that had previously been collected by the same individuals, although for a different purpose than their re-use to explore new research questions (Ruggiano and Perry, 2019).

From a much broader perspective, secondary research can also extend beyond the re-use or re-analysis of data from previous primary research, although the purpose of the data's existence should be different from the purpose of their present use in research. Secondary research can encompass the analysis of existing datasets that might not have come about from primary research (Doolan *et al.*, 2017). Doolan *et al.* (2017), whose work was done from a medical research perspective, emphasised that existing datasets for secondary research can also be derived from other sources such as hospital charts, academic course records, quality improvement records, news media, or social media. What is clear from the various perspectives is that secondary research: (1) involves the use of pre-existing datasets, and (2) these datasets should have come into existence for purposes that are different from the purpose of their use in the secondary research. This view is consistent with the meaning of the term "secondary" in secondary research, which refers to the use of pre-existing data for secondary analysis. The term indicates only that the data are being used for research purposes beyond the specific need that prompted their original gathering or its generation (Stewart and Kamins, 1993).

The term "research" refers to the application of a systematic approach to study and generate new facts and conclusions about a subject of interest. Put together, a broader definition of "secondary research" is the study of specific problems to generate new facts and conclusions through analysis of pre-existing data or information that was originally created for a different reason or purpose.

It is from this broader perspective of secondary research that the rest of this book has been compiled. Based on this broader perspective, it should be emphasised once again that the pre-existing data or information may or may not have come about from primary research but should still meet the requirement of existing for a different purpose than their use in secondary research. The systematic approach that should be applied when conducting secondary research requires that a clear, logical, transparent, and verifiable process is followed.

Secondary research process

Despite the increase in the use of secondary research methods, there is still limited guidance on what this process should entail (Doolan *et al.*, 2017), and practical case examples that outline the process and techniques required to carry out secondary research effectively are lacking, particularly within the context of the built environment. However, it should be clarified that one of the main features that differentiates the secondary research process from the primary research process relates to data collection and analysis as shown in Table 1.1.

The steps involved in the secondary research process are discussed briefly below.

Table 1.1 Comparison of secondary data and primary data research processes

Secondary research	Primary research
Establish gaps in the research and formulate research questions using existing literature	Establish gaps in the research and formulate research questions using existing literature
Undertake a detailed literature review	Undertake a detailed literature review
Identify, select, and evaluate the existing datasets	Develop data collection instruments and protocol and collect data
Undertake secondary data analysis	Undertake primary data analysis
Discuss, interpret, and disseminate findings from the research	Discuss, interpret, and disseminate findings from the research

Source: Adapted from Doolan *et al.* (2017) and Johnston (2014).

Establish gaps in the research and formulate research questions

Just like with primary research, it is important that secondary research commences with a literature review to establish the research gaps, and research question(s) or hypotheses. Clearly defined research question(s) or hypotheses will be crucial in establishing whether the study will fit well with any existing dataset (Smith, 2008).

Undertake literature review

At this stage of the secondary research process, literature will need to be reviewed to examine the current and past thoughts and issues in the area of interest (Johnson, 2014). This is still the case even when the secondary method to be used involves the re-analysis of published academic literature as secondary data e.g. a systematic literature review.

Identify, select, and evaluate the existing datasets

It is important to ascertain whether the existing dataset can address the research question of the secondary study (Long-Sutehall *et al.*, 2010; Johnson, 2014). If the datasets are from previous primary research, it is essential that the purpose for which they were collected originally – the data collection techniques and instruments used and the participants from which the data were collected – are all established as part of the evaluation (Smith, 2008).

Undertake secondary data analysis and disseminate findings

Data analysis forms an integral aspect of any research methodology. It is essential that the analysis and the interpretation of the findings are undertaken in the same way as the methods used for the primary data research (Long-Sutehall *et al.*, 2010).

Secondary research designs

Just as in primary research, secondary research designs can be either quantitative, qualitative, or a mixture of both strategies of inquiry (qualitative and quantitative) as shown in Figure 1.1. Qualitative secondary research designs can be based on either the re-analysis of published results or the analysis of existing qualitative datasets. Largan and Morris (2019, p. 14) defined qualitative secondary research as "a systematic approach to the use of existing data to provide ways of understanding that may be additional to or different from the data's original purpose". There are various qualitative, secondary research methods. Those that are based on the re-analysis of published academic literature include SLRs or meta-synthesis, state-of-the-art reviews and scoping reviews. SLRs have been discussed in detail in Chapters 5–7 of this book. Qualitative secondary research can also be based on

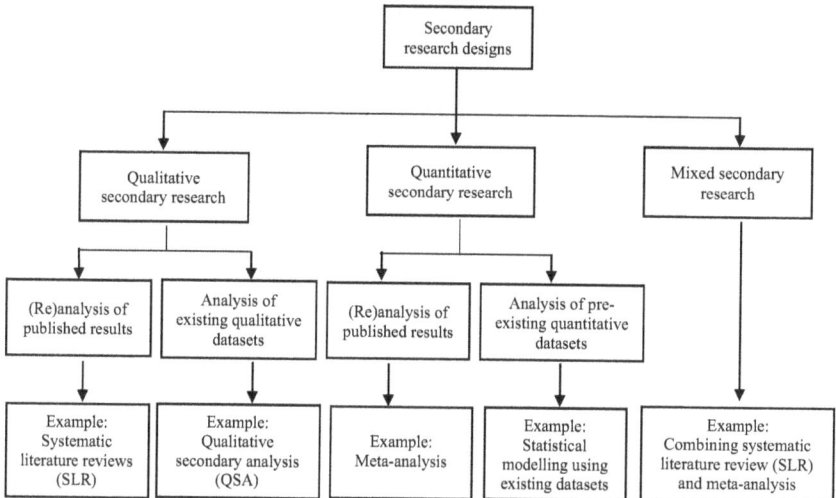

Figure 1.1 Secondary research designs.
Source: Original.

designs that involve the analysis of existing raw qualitative datasets rather than the published results from a previous analysis. A typical example of such a design is the QSA, which involves the use of pre-existing qualitative data from primary research to develop new social scientific or methodological understandings (Heaton, 2008; Irwin, 2013). According to Johnston (2014), using QSA contributes to scientific knowledge by offering alternative theoretical or conceptual perspectives on previously collected and archived qualitative data. QSA has been discussed in detail in Chapter 4 of this book.

There are other qualitative designs like legal research that arguably straddle between the primary and secondary research domain, depending on whether the sources of data used for the analysis are regarded as primary or secondary sources of the law. For example, the analysis of legal trends and principles using secondary sources of law, such as law reports, legal commentaries, and other literature about law, will constitute a form of secondary qualitative research. The secondary dimension of legal research, and more specifically doctrinal legal research, has been discussed in detail in Chapter 8 of this book.

Similarly, on the quantitative secondary research side, designs can involve the re-analysis of published academic literature or the analysis of pre-existing quantitative datasets. A typical example of a quantitative secondary research design that is based on published results is quantitative secondary analysis, which is a quantitative form of systematic review that is commonly referred to as meta-analysis research. This is a very well-established secondary research design in which statistical approaches are used to combine quantitative research findings from multiple empirical studies to increase the analytical power owing to the combined effect of sample sizes from the various studies. Church (2001) described meta-analysis as the quantitative combination of statistical information from multiple studies on a given phenomenon. Meta-analysis has been applied in Chapter 15 of this book.

Another secondary design that is more quantitative in nature and which is growing in popularity in the built environment is bibliometric research. Bibliometric research or analysis and its sub-approaches, such as scientometric research or science mapping, are alternative quantitative approaches to SLRs and are used to synthesise various trends from published academic literature. Bibliometric research involves the analysis of quantitative patterns relating to a cluster of scientific documents within a given domain (De Bellis, 2009) through the computation of quantitative metrics from published academic literature. The analytical methods that are used in bibliometric research include co-author analysis, co-citation analysis, and keyword analysis. Bibliometric research and its sub-approaches, such as scientometric research, have been presented in Chapters 9–13, including case examples that reflect their growing popularity across various built environment disciplines.

Literature-based discovery (LBD) is another less known method also on the quantitative spectrum that can be used to synthesise results from published academic literature. This method involves the use of statistical procedures to deduce relationships and hypothesis from published academic literature by computing semantic measures. LBD has been discussed in detail in Chapter 14.

The secondary quantitative research designs that are based on analysis of pre-existing quantitative datasets involve the use of statistical techniques to analyse different trends in the data. Such techniques include statistical modelling and the application of big data analysis techniques to existing datasets. These techniques could also involve the re-use of quantitative data from previous primary research by applying different statistical analysis to generate different insights.

Secondary research can be based on a mixed design, whereby a mixture of secondary qualitative and secondary quantitative approaches is applied in a study. For example, this could be the application of SLRs and bibliometric reviews or SLRs and meta-analysis, the latter of which has been demonstrated in Chapter 15 of this book. In Chapter 15, Low-Choy *et al.* advocate a mixed approach to meta-analysis research (qualitative and quantitative). An example of a mixed secondary design that utilises existing quantitative and qualitative datasets is also presented in Chapter 16. This chapter is used to demonstrate the potential in using existing datasets, collected as part of previous research, to answer new research questions, having ensured the suitability of the existing data for the new studies. It should be noted that the secondary research designs discussed in this chapter are not exhaustive, but that the focus has been mainly on methods that have been covered in this book.

It is evident from the discussions and from Figure 1.2 that the process that should be followed when designing secondary research is similar to that of primary research. This decision-making process involves a reflection on the philosophical assumptions underpinning the study (positivist or interpretivist), which should inform the

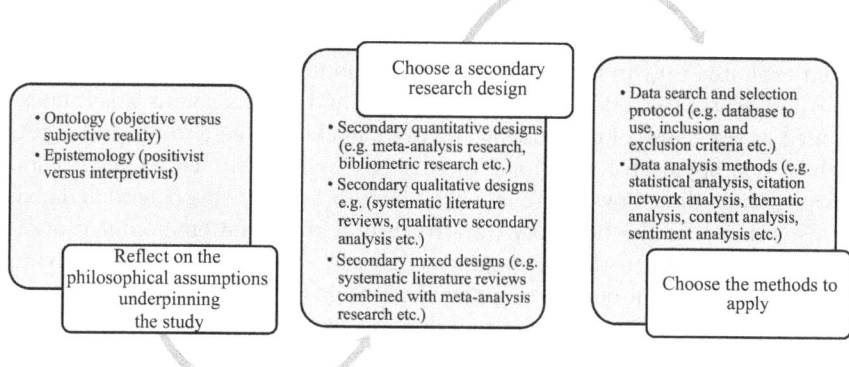

Figure 1.2 Framework to guide decision-making when designing secondary research.
Source: Original.

choice of the research design (meta-analysis research, bibliometric research, SLR and so on) and the research methods (thematic analysis, content analysis, citation network analysis and so on). Although these philosophical considerations might be hidden in the decision-making process, they still influence the choices that are made in terms of the research strategy and the methods that are adopted for organising and analysing the secondary data. As an example, in a SLR study that is aligned to an interpretivist perspective, thematic analysis would be selected as the method of data analysis as opposed to content analysis (see Chapter 6).

Potential for applying secondary research methods in the built environment

There is considerable potential to apply both qualitative and quantitative secondary research methods across various built environment disciplines. Research students in the fields of construction management, construction project management, quantity surveying, construction law and dispute resolution, real estate and property management, and civil engineering can apply secondary qualitative or quantitative methods that are based on either re-analysis of published academic literature or re-use of pre-existing raw datasets.

To date, there has been a significant increase in the use of methods that are based on published academic literature, such as SLRs, and bibliometric research. There have also been some applications of meta-analysis research in built environment research. For example, Alruqi and Hallowell (2019) performed a statistical meta-analysis of leading indicators of construction safety by combining the results of various studies to increase the magnitude of effect and significance of the findings. However, there is scope for increased use of meta-analysis research as a quantitative approach for synthesising results from already published results. There is also potential for the application of secondary methods such as scoping reviews, state-of-the-art reviews and LBD, which are also based on the re-analysis of published academic literature, but have not been exploited fully in built environment research.

Another area that offers significant potential, and which remains largely unexploited, is the re-use of pre-existing, archived, qualitative data to conduct QSA. Whilst this approach has become increasingly well established in mainstream social science disciplines, there is a need for awareness of this potential. There is also a need for the necessary infrastructure within built environment disciplines such as construction management that can enable sharing and archiving of original data from qualitative primary research so that these can be re-used as secondary data. Opportunities to utilise pre-existing qualitative datasets (e.g. social media data) and quantitative datasets that are not necessarily a result of previous primary research will also continue to grow with the increasing availability of such datasets and the advancements in data mining and big data analytic techniques that make it easier to explore trends in these datasets. Despite the potential and advantages of utilising secondary research methods in the built environment, there are also potential challenges and drawbacks.

Benefits and drawbacks of applying secondary research methods

Secondary research methods can offer a range of benefits in built environment research, but are not without drawbacks. The benefits of using secondary research methods include the following. The methods:

- are comparatively quicker than primary research and can save time;
- are less expensive than primary research;
- can offer possibilities to study topics that are too sensitive to undertake by engaging first-hand with people and institutions;
- can enhance the scale of research that can be conducted even with time and resource constraints;
- can help to prevent respondent fatigue, which is a usual occurrence with primary data collection as participants become tired of completing questionnaire surveys or participating in interviews.

The earlier advantages reflect the views of Johnston (2014), which are that secondary research can be quicker, more cost-effective, and convenient, allowing for analysis of larger datasets that are more representative of the target population, and ensuring that higher levels of validity and more generalisable findings are achieved. However, it should be noted that, in some instances, secondary data might be available at a cost (e.g. on subscription basis). Also, contrary to the view that using secondary research can be quicker, it could take a considerable amount of time to transform into a format required for the analysis. This raises some of the challenges and drawbacks associated with the application of secondary research methods for built environment research, which include the following:

- The data might not be available in the format required for analysis, thus requiring considerable time and effort to transform;
- The available data might not necessarily be appropriate for answering the research questions of interest for the secondary research;
- Methodological challenges could result from the use of inappropriate data analysis methods, leading to misinterpretation of the original data;
- There also might not be enough contextual information about the data to enable accurate interpretation of the findings.
- The realities at the time when the original data came into existence might not be the same as the realities during the period of secondary analysis, raising an issue about the timeliness of the data;
- In instances where the secondary research utilises data from previous, primary research, the secondary researcher might not be privy to the problems or weaknesses associated with the collection of the primary data as, in most instances, they would not have been part of that process;
- There might be missing sections in the existing datasets (e.g. missing datasets in time-series data) that might be difficult to extrapolate with accuracy.

Despite the potential as well as considerable opportunities for applying secondary research methods in built environment research, it is evident from the challenges and drawbacks listed earlier that there are also pitfalls and limitations that should be considered carefully. The application of secondary research methods might not necessarily be appropriate for the research questions of interest. It will be necessary for built environment researchers, who are interested in applying secondary research methods, to be aware of these pitfalls and limitations so that they can make the best decisions about whether to apply secondary research methods in the first place.

Overview of chapters

This book comprises 16 chapters which are outlined below.

In Chapter 1, Introduction to Secondary Research Methods in the Built Environment, Emmanuel Manu and Julius Akotia present discussions on what constitutes secondary research before reflecting on the secondary research process, secondary research designs, and the potential for applying secondary research methods in built environment research. The benefits and drawbacks of applying secondary research methods also have been reflected upon. The purpose of this chapter was to provide the context for all the other chapters in the book.

In Chapter 2, Identifying and Sourcing Data for Secondary Research, Emmanuel Manu, Julius Akotia, and Saad Sarhan distinguish between secondary data and primary data, whilst acknowledging the sometimes blurred and confusing nature of this distinction. The sources of secondary data that are applicable to various built environment disciplines have also been identified and discussed. These include academic databases, government databases, intergovernmental databases, organisational databases, legal databases, and social media data. The opportunities for utilising data from these various sources have been interrogated. Manu *et al.* also reflect on some of the considerations that will need to be exercised to ensure that the pre-existing datasets are of the right quality for use in secondary research.

In Chapter 3, Ethical Considerations in the Use of Secondary Data for Built Environment Research, Abid Hasan evaluates the ethical considerations that must be exercised when using secondary data for research. In this chapter, the main ethical issues that must be considered as part of the research design – data collection and analysis, data storage and disposal, and dissemination of the findings – when using secondary datasets are addressed. Just as with primary research, Hasan makes it clear that the research protocol for secondary data research should address the questions regarding how the existing data will be collected, analysed, kept anonymous, published, stored, and secured. By drawing on the three Belmont principles of autonomy (respect for persons), beneficence, and justice, Hasan ends with advice that researchers using secondary data should ensure that sampling protocols and data collection procedures are established, the source and ownership of the original data is acknowledged, information about the original data (e.g. the response rate, sampling bias, missing data, and the time data came into existence) is reported, and that effective methods of securing the anonymity of the data are used to report the findings of the secondary research.

In Chapter 4, Qualitative Secondary Analysis as a Research Methodology, Victoria Sherif provides a historical overview of QSA as a research methodology. Epistemologically, Sherif highlights the systematic, subjective and yet highly reflexive process of QSA as the researcher explores pre-existing qualitative data for new meanings relating to human experiences within a social context. This reflexive process, which should account for the background context of the original data, has the potential to generate more meaningful empirical and/or methodological findings.

Methodologically, Sherif highlights the extensive preparatory and evaluative work that is required to address the research objectives adequately and meaningfully when applying QSA. It is also advised that QSA should be applied as a methodological approach for generating new knowledge or broadening understanding of a topic of interest, to enable a new research study or data collection or to use as a discrete method. Sherif also advises that the fit and relevance of the dataset, general quality of the dataset, trustworthiness of the dataset, and timeliness of the dataset are considerations that should be exercised before selecting existing data for QSA. Whilst this chapter is focused on the social and educational context, it offers insight into the application of QSA as a methodology in built environment research.

In Chapter 5, Evaluation of Systematic Literature Reviews in Built Environment Research: What are We Doing and How Can We Improve?, Vijayan Chelliah, Nicola Thounaojam, Ganesh Devkar, and Boeing Laishram introduce the SLR technique by reflecting on five primary steps of the SLR process comprising: formulate the research question, locate the literature, select and evaluate the literature, analyse and synthesise the studies, and report the review results. They offer important advice on how to formulate concise research questions for SLRs using different frameworks, how to locate the literature using appropriate search strings, the inclusion and exclusion criteria for selecting and evaluating the literature, some standardised questions for analysing and synthesising the findings, and approaches and procedures for reporting the results. Chelliah *et al.* also evaluate how these five stages of SLR have been applied in built environment research, based on which they have identified areas of improvement. Based on this evaluation, they have provided a very useful checklist for improving the quality of SLR studies in the built environment.

In Chapter 6, When Does Published Literature Constitute Data for Secondary Research and How Should the Data be Analysed?, Saad Sarhan and Emmanuel Manu discuss the use of published academic literature for both traditional literature reviews in research as well as for secondary data when applying qualitative secondary research methods, such as SLRs, scoping reviews, state-of-the-art reviews, or quantitative secondary research methods such as bibliometric reviews and meta-analysis. Sarhan and Manu then focus on the details of SLR as a qualitative secondary research method, based on which they discuss thematic analysis and qualitative content analysis as the main methods for analysing qualitative secondary data, with a focus on using computer-assisted, qualitative data analysis software such as NVivo to support this process. Specific guidance on when and how to use NVivo for supporting qualitative SLRs is also presented. This chapter, therefore, contains methodical guidance on how to conduct SLRs of existing academic literature, using NVivo – an approach which can also be applied to other qualitative secondary research methods.

In Chapter 7, A Systematic Literature Review Evaluating Sustainable Energy Growth in Qatar Using the PICO Model, Redouane Sarrakh, Suresh Renukaappa, and Subashini Suresh illustrate the use of the PICO Model for SLRs with an example of a study that evaluates sustainable energy growth in Qatar. This chapter includes a case example of the application of SLRs following the PICO Model in built environment research, based on a case study about the efficiency of policies and tactics implemented by the Qatari Government in its energy sector, pertaining to sustainability strategies. Initial results using the PICO Model led to the identification of 1990 resources from five different databases, of which 82 met the pre-set inclusion and exclusion criteria, such as date, geographic location, language, type of publications, participants, and design studies. From the SLRs, Sarrakh *et al.* were able to map the Qatar Energy Sector to six sustainability initiatives, namely: health and safety, environment, climate change and energy, economic performance, society, and workforce. Sarrakh *et al.* concluded that Qatari sustainable development policies still needed great efforts to achieve more holistic policies and more integrated and comprehensive strategies.

In Chapter 8, Understanding Legal Research in the Built Environment, Joseph Mante discusses legal research as an approach that employs both primary and secondary sources of data to arrive at logically sound outcomes. Mante argues that within the built environment, legal research is often undervalued or even mis-characterised as a tool for preliminary enquiry. These misconceptions stem from lack of understanding of the province of legal research in the built environment and the procedures involved. This misunderstanding is dispelled in this chapter by explaining the scope and the procedures involved in legal research, with doctrinal legal research being a dominant aspect. In its basic form, legal research involves locating, describing, interpreting, and systematising legal principles and concepts, with the legal system as a conceptual framework. The resources for this exercise are primary data (legislations) and secondary data (e.g. law reports, legal commentaries, and other literature about the law), and the outcomes are supported and based on sound reasoning.

In Chapter 9, Applying Science Mapping in Built Environment Research, Amos Darko and Albert Ping-Chuen Chan discuss science mapping as an effective and useful methodology for studying and understanding the structural and dynamic features of a scientific domain through constructing, analysing, and visualising bibliometric networks. Darko and Chan discuss the application of science mapping in built environment research before providing a step-by-step tutorial on how three software packages, VOSviewer, CiteSpace, and Gephi, can be applied together to conduct robust science mapping-based research. This chapter will be helpful to researchers and other interested stakeholders that intend to undertake quality research using science mapping.

In Chapter 10, Bibliometric Analysis for Reviewing Published Studies in the Built Environment, Liyuan Wang, Ruoyn Jin, and Joseph Kangwa define bibliometrics analysis before reflecting on the rationale for adopting this method when conducting literature reviews. The existing software tools for conducting the text mining-based analysis (e.g. VOSviewer, Gephi) also are introduced. Using two case examples from disciplines in the built environment, Wang *et al.* illustrate the science mapping workflow that is involved in bibliometric analysis, based on which Wang *et al.* showcase the network analysis with one of the bibliometric analysis tools (i.e. VOSviewer). Finally, general guidance that should be observed when conducting bibliometric analysis is provided, with some concluding recommendations on the common mistakes that should be avoided when conducting bibliometric analysis.

In Chapter 11, Scientometric Review and Analysis: A Case Example of Smart Buildings and Smart Cities, Timothy O. Olawumi, Abdullahi B. Saka, Daniel W.M. Chan, and Nimesha S. Jayasena present the scientometric analysis process as a quantitative study of the intellectual evolution of research themes based on large-scale datasets before presenting a case example of this method using a study on smart buildings and smart cities. Using this example, Olawumi *et al.* reflect on simplified steps that should be followed when conducting scientometric analysis, addressing issues such as sources of data for the "smart buildings and smart cities" research theme, software tools that can be utilised, and the analysis that can be performed to identify trends using the citation data such as analysis of co-author network, co-occurring keywords, author co-citation network, and document co-citation network. The case example serves as a useful guide for built environment researchers who are interested in applying scientometric analysis to other emerging research themes.

In Chapter 12, Analysis of BIM-FM Integration Using a Science Mapping Approach, Ecem Tezel and Heyecan Giritli present another case example on science mapping, using a study on building information modelling (BIM) and facility management (FM). Through this case example, they demonstrate the application of a three-step, science mapping approach to the BIM and FM knowledge domain. These three steps comprise a bibliometric search of the journal articles published in the Web of Science and Scopus databases, followed by scientometric analyses of the journals using VOSviewer software to identify the most influential journals, authors, and keywords in the BIM-FM domain, before finalising the third step, which is an in-depth qualitative discussion to summarise the present knowledge in BIM-FM integration and to propose future research directions. Through this case example, Tezel and Giritli demonstrate the application of three analytical domains for science mapping studies, namely journal analysis, scholar analysis, and keyword analysis.

In Chapter 13, Trends in Recycled Concrete Research: A Bibliometric Analysis, Olalikan Shamsideen Oshodi and Bankole Osita Awuzie present a case example of a bibliometric study. The aim is to detail the research trends and gaps associated with material circularity of concrete to identify knowledge gaps and future research directions. The topic of material circularity has been attracting attention from the construction industry stakeholders that are keen to overcome the industry's negative impacts on sustainability and, since concrete is an intensively utilized resource in the construction industry, Oshodi and Awuzie chose a bibliometric analysis as a method for establishing the research trends on circularity of concrete as a construction material, whilst highlighting gaps necessitating further study using recycled concrete. This bibliometric method allowed for the identification of the growth in rate of publications within the review period, the most productive authors working within the knowledge domain area, degree of collaboration between them, author distribution, collaboration networks, institutions and countries producing such publications, and the journals where such articles were published. The example presented in this chapter can provide guidance for construction management researchers who are interested in applying bibliometric analysis.

In Chapter 14, Using Literature-Based Discovery in Built Environment Research, Nathan Kibuwami and Apollo Tutusigensi introduce a secondary research method called LBD, which involves the identification of novel relationships and/ or theories from two or more disparate contexts of literature. With origins in biomedical research, LBD is used to search for novel hypotheses in the literature, using either an open discovery or closed discovery approach. Kibuwami and Tutusigensi argue that there has been very limited application of LBD in built environment research despite the potential it offers, with some built environment researchers apparently confusing this method with other literature-related approaches such as SLRs. Kibuwami and Tutusigensi continue to advocate the development of a robust understanding of LBD among built environment researchers in order to increase its use. They achieve this by proposing a five-step approach to implementing LBD, which involves literature data retrieval; term extraction; category development; semantic similarity; and deduction of relationships. This five-step approach is applied using a case example to demonstrate how the core principles of LBD can be upheld.

In Chapter 15, Combining Study Findings by Using a Multiple Literature Review Technique and Meta-Analysis: A Mixed Method Approach, Samantha Low-Choy, Fernando Almeida, and Judy Rose discuss the meta-analysis research process before presenting two inter-disciplinary case examples of meta-analysis research. They adopt a mixed approach to meta-analysis research that commences with a structured literature review (scoping and then SLR) to select studies that are used to perform the meta-analysis research, clarifying eligibility via qualitative, narrative or model-centric review, and ending with a realist review. The AMSTAR2 appraisal tool is applied to a seven-staged process for conducting meta-analysis studies, which Low-Choy *et al.* then apply to two, non-randomised case studies. The guide they provide to performing meta-analysis research is based on AMSTAR2, which is a meta-analysis appraisal tool for appraising the quality of published systematic reviews for meta-analysis. Low-Choy *et al.* suggest that viewing meta-analysis as a mixed (quantitative and qualitative) method provides a wider array of options and is more suitable in many fields, especially multi-disciplinary fields such as built environment. Through these two case examples, Low-Choy *et al.* provide a forward-looking guide for researchers who will be interested in conducting meta-analysis research using observational data, which is the type of data that is prevalent within built environment disciplines such as construction management.

In Chapter 16, Analysing Secondary Data to Understand the Socio-Technical Complexities of Design Decision Making, Payam Pirzadeh, Helen Lingard, and Nick Blismas present a secondary research study that involved the selection and re-analysis of six case studies from an existing comprehensive dataset. The existing dataset included 23 case studies, each of which was focused on the building-design process of a structural element. This existing dataset had come about from a study that had a different purpose of understanding in which characteristics of communication between participants in the design process were linked to positive health and safety (H&S) outcomes. The aim of the new, secondary research study was to reveal the interdependence between social and technical aspects of construction-design decision-making and explain the impact on constructability and H&S outcomes by building on and extending the findings of the previous research. To select six of the existing cases for re-analysis, Pirzadeh *et al.* developed and used a set of selection criteria to ensure the suitability of the secondary data for the new study. Pirzadeh *et al.* then applied a secondary, convergent, mixed methods design that was combined with a novel, multi-level network analysis framework to integrate and analyse the existing, quantitative, and qualitative data for each case simultaneously. Consequently, a more comprehensive and detailed investigation of the socio-technical complexities that characterise construction-design decision-making was achieved in the new study. To demonstrate this approach, the results of only one of the case studies have been reported in Chapter 16. Opportunities for employing new research designs and novel methods to re-analyse existing datasets, collected as part of previous research, to answer new research questions are indicated in the chapter. The importance also of ensuring the suitability of the existing data for new studies is highlighted.

References

Alruqi, W.M. and Hallowell, M.R. (2019). Critical success factors for construction safety: Review and meta-analysis of safety leading indicators. *Journal of Construction Engineering and Management*, 145(3), pp. 1–11.

Church, R.M. (2001). The effective use of secondary data. *Learning and Motivation*, 33, pp. 32–45. doi:10.1006/lmot.2001.1098.

De Bellis, N. (2009). *Bibliometrics and Citation Analysis: From the Science Citation Index to Cybermetrics*, Scarecrow Press, Plymouth.

Doolan, D.M., Winters, J. and Nouredini, S. (2017). Answering research questions using an existing data set. *Medical Research Archives*, 5(9), pp. 1–14.

Driscoll, D.L. (2011). Introduction to primary research: Observations, surveys and interviews. *Writing Spaces: Readings on Writing*, 2, pp. 153–174.

Heaton, J. (2008). Secondary analysis of qualitative data: An overview. *Historical Social Research*, 33(3), pp. 33–45. doi:10.12759/hsr.33.2008.3.33-45.

Irwin, S. (2013). Qualitative secondary data analysis: Ethics, epistemology and context. *Progress in Development Studies*, 13(4), pp. 295–306. doi:10.1177/1464993413490479.

Johnston, M.P. (2014). Secondary data analysis: A method of which the time has come. *Qualitative and Quantitative Methods in Libraries*, 3, pp. 619–626.

Largan, C. and Morris, T. (2019). *Qualitative Secondary Research: A Step-by-Step-Guide*, Sage, Thousand Oaks, CA.

Long-Sutehall, T., Sque, M. and Addington-Hall, J. (2010). Secondary analysis of qualitative data: A valuable method for exploring sensitive issues with an elusive population? *Journal of Research in Nursing*, 16(4), pp. 335–344. doi:10.1177/1744987110381553.

Ruggiano, N. and Perry, T.E. (2019). Conducting secondary analysis of qualitative data: Should we, can we, and how? *Qualitative Social Work*, 18(1), pp. 81–97. doi:10.1177/1473325017700701.

Smith, E. (2008). Pitfalls and promises: The use of secondary data analysis in educational research. *British Journal of Education Studies*, 56(3), pp. 323–339. doi:10.1111/j.1467–8527.2008.00405.x.

Stewart, D. and Kamins, M. (1993). *Secondary Research: Information Sources and Methods*, Second Edition, Sage Publications, Newbury Park, California

2 Identifying and sourcing data for secondary research

Emmanuel Manu, Julius Akotia, Saad Sarhan, and Abdul-Majeed Mahamadu

Introduction

We live in a world full of data, with advances in digital technologies making it possible for data to be extracted from various sources for secondary research. Government agencies and other international organisations have adopted open data initiatives, making their data freely available for public use (Gebre and Morales, 2020). Technological advancement has also made it possible for government agencies and research organisations to store research data so that they are easily and readily available for re-use by other researchers (Irwin and Winterton, 2012). Several organisations provide access to pre-existing data about a broad spectrum of issues either freely or on a subscription basis. Opportunities for data mining and analysis have also emerged because of the rapid increase in the use of data from social media. This has made it possible for opinions and interactions among complex networks of individuals to become accessible, searchable, and exploitable for research purposes amongst social scientists (Felt, 2016). With this opportunity to exploit a vast amount of data from various sources for secondary research, Tight (2019, p. 14) stated that "It will be foolish not to at least explore the possibilities [of using secondary data], whether as part of a research project or as the whole project before committing to further data collection". It is in recognition of this potential that the aim of this chapter was to identify and discuss the potential sources of data for secondary research in the built environment. The chapter begins with discussions on what constitutes secondary data, before identifying and discussing some of the sources of secondary data and the issues about the quality of secondary data that should be addressed before being applied for the purposes of secondary research.

What constitutes secondary data?

It is not always possible to arrive at clear distinctions between primary and secondary data (Smith, 2008; Largan and Morris, 2019). However, Doolan *et al.* (2017, p. 2) have defined a dataset as "any set of existing data that could be used to answer important new research questions and/or provide further evidence relevant to ongoing research questions". Secondary data are described also as facts and information that exist for some purpose other than the purpose of the immediate study at hand (Rabianski, 2003), while primary data are facts and information collected specifically for the purpose of an investigation at hand. This suggests that primary data are those that are gathered first-hand for a study, whereas secondary data are pre-existing data that are compiled by the analyst from either published or unpublished work. It is important to recognise that secondary data are sourced not only from published, scholarly work, but also unpublished work. Doolan *et al.* (2017) emphasised that pre-existing datasets either could be derived from previous research or could be existing data from other sources such as hospital charts, academic course records, quality improvement records, news media, or social media. According to Smith (2008), secondary data can be captured in a broad spectrum of empirical forms, existing as either numeric or non-numeric data, with non-numeric secondary data also including those that are retrieved second-hand from interviews, ethnographic accounts, documents, photographs, or conversations. It is for these reasons that Rabianski (2003) differentiated between primary sources of secondary data and secondary sources of secondary data. Rabianski, (2003) described primary sources of secondary data as being original data in its raw state that has been archived for re-use, e.g. census data and other research data, whereas secondary sources of secondary data are sources that make data available after it has been processed, i.e. manipulated, augmented, modified, summarised, or synthesised.

Sources of secondary data

The application of secondary research methods is dependent on the identification and sourcing of pre-existing datasets that are appropriate for addressing the research question(s) of the secondary study. Secondary data could be derived from primary or secondary sources, following the classification by Rabianski (2003). Primary sources of secondary data will include, for example, archived transcripts of interviews conducted by primary researchers and data from social media. Secondary sources of secondary data will include published scholarly articles, books, government documents, reports, official historical records, statistical databases, and administrative records of organisations. These datasets are accessible from a range of sources which have been classified in Figure 2.1 as comprising: academic databases, government databases, databases of intergovernmental agencies, organisational databases, social media data, and legal databases. Some examples of these databases have been summarised in Table 2.1. It is also important to note that Google has launched a dataset search engine (https://datasetsearch.research.google.com/) that can be used to access over 25 million datasets on various subjects. The search engine links the user to different research datasets and can be filtered according to type of data, e.g. images, tables, or text.

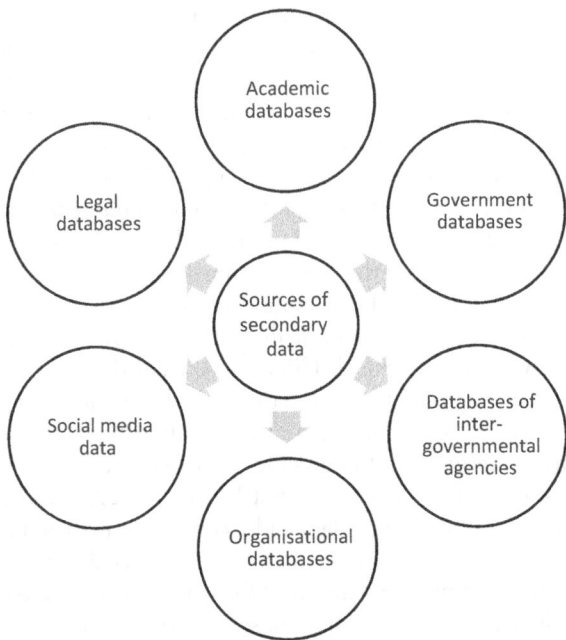

Figure 2.1 Sources of secondary data.

Academic databases

Secondary data from academic and scholarly databases can take the form of either secondary, published, scholarly work, or primary datasets from past research. The published, scholarly data can be used when applying secondary research methods, such as systematic literature reviews, scoping reviews, meta-analysis, and bibliometric and scientometric research. The scholarly data can be retrieved from academic and scholarly databases such as Web of Science, Scopus, Google Scholar, and ProQuest, among others, through search queries that have been constructed carefully based on the topic of interest (see Chapter 5).

There are also academic databases that contain an archive of raw datasets from past research projects. Numerous repositories and archives have been established (Bishop and Kuula-Luumi, 2017), which has led to the availability of, and access to, existing datasets from previous primary research (Sherif, 2018). This practice has been driven by an increase in official policies that promote and facilitate the retention, sharing, and re-use of qualitative datasets in research (Heaton, 2008). There have been demands also from an increasing number of funding institutions, professional associations, and journal editors that authors make their data publicly available for re-use by secondary researchers

Table 2.1 Some examples of sources of secondary data for built environment research

Category of database	Name of database	Web address	Type of data
Academic databases	Web of Science	www.webofknowledge.com	Scholarly data
	Scopus	https://www.scopus.com/sources	Scholarly data
	ProQuest	https://www.proquest.com/	Scholarly data
	Google Scholar	https://scholar.google.com/	Scholarly data
	UK Data Service	https://www.data-archive.ac.uk/	Archived raw data from previous research
Government databases	UK Government Database	www.data.gov.uk	Variety of datasets
	US Government Database	https://www.data.gov	Variety of datasets
Inter-governmental agency database	Eurostat Database	https://ec.europa.eu/eurostat/	EU statistical data
	EU Data Portal	https://www.europeandataportal.eu/en	Variety of datasets
Organisational databases	FAME Database	https://www.bvdinfo.com/en-gb/our-products/data/national/fame	Financial data on public and private companies
	World Bank Database	https://data.worldbank.org/	Variety of datasets
	GEMS/Water Data Centre	https://gemstat.org/	Global freshwater datasets
	Ecoinvent Database	https://www.ecoinvent.org	Material and product data for life-cycle inventory
Legal database	Westlaw	https://legalsolutions.thomsonreuters.co.uk/en/products-services/westlaw-uk.html	Legal datasets
	LexisNexis	https://www.lexisnexis.com/uk/legal/	Legal datasets
	British and Irish Legal Information Institute	https://www.bailii.org/	Legal datasets

Source: Original.

(Bishop and Kuula-Luumi, 2017; Feldman and Shaw, 2019). This policy direction has advanced the increase and application of secondary data by many researchers (Greenhoot and Dowsett, 2012). In the United Kingdom, for example, there are requirements that applicants seeking the Economic and Social

Research Council (ESRC) grants provide justification for the need to collect new data for their research projects. Applicants are also required to provide information on whether they have explored the potential of existing datasets adequately to establish whether these are inadequate for their proposed study (Morrow *et al.*, 2014). Additionally, the ESRC has commissioned the Secondary Data Analysis Initiative, whereby funding is awarded for projects that are dedicated to the use of secondary data. Through these efforts, qualitative and mixed methods datasets are now disseminated by the UK Data Archive, which is part of the larger UK Data Service, through their ReShare portal (reshare. ukdataservice.ac.uk) (Bishop and Kuula-Luumi, 2017). Some journal publishers have also taken similar steps by requesting that original data are retained and shared for future re-use (Church, 2001).

Whilst it is a growing practice within the mainstream social science disciplines, the secondary analysis of archived raw datasets from past research projects has not become a common practice yet within built environment disciplines such as construction management. Thus, there is opportunity to establish infrastructure to facilitate the retention, sharing, and re-use of qualitative datasets either by journal publishers adjusting editorial policy in the built environment or by making it a requirement for funded projects.

Government databases

Many countries around the world now operate open data initiatives, ensuring that public sector data is organised, preserved, and made freely available for public use through government and public sector institutions (Gebre and Morales, 2020). These datasets are accessible online through open data portals. For example, the UK Government's open data can be accessed at: https://data.gov.uk, the US Government's at: https://www.data.gov, the Canadian Government's at: https://open.canada.ca, the Indian Government's at: www.data.gov.in, and the open data of various African countries at: https://africaopendata.org and https:// dataportal.opendataforafrica.org. Platforms such as https://dataportals.org, run by the Open Knowledge Foundation, also provide a comprehensive list of open data portals from governments around the world. These publicly available open datasets, sometimes referred to as "Open Government Data" (Zuiderwijk *et al.*, 2016), are intended to facilitate civic engagement and government transparency (Gebre and Morales, 2020) and have implications for secondary research in the built environment. For example, sets containing data about various relevant matters, such as air quality, the impact of built environment-related policies, national statistics on property transactions, size and shape of the construction industry, construction waste, public construction projects, and national performance of built environment sectors, are all accessible through these government open data portals and can be used for secondary research in the built environment.

Databases of inter-governmental agencies

These databases are similar to government databases, but provide data relating to regions rather than from one government. Most of these inter-governmental agencies also operate open data initiatives. For example, inter-governmental agencies such as the EU operate an open data initiative through the EU Open Data Portal (https://www.europeandataportal.eu/en) providing a single point of access to data captured by EU institutions, agencies, and bodies for anyone to re-use. Some of these datasets are relevant and suitable for secondary research in the built environment.

Organisational databases

There are many private and public sector organisations that provide access to datasets either on a subscription basis or as open data. Examples of organisational databases include the FAME database (https://www.bvdinfo.com/en-gb/our-products/data/national/fame) as a source of financial information for more than two million public and private companies in the United Kingdom and Ireland, which can be applied in the financial analysis of construction organisations. The Building Cost Information Service Database (https://service.bcis.co.uk/), run by the Royal Institution of Chartered Surveyors, is a source of data about building costs and tender prices, output indices, and other construction statistics that can be applied to study trends in building cost. The Ecoinvent Database (https://www.ecoinvent.org/) is another example, providing a source of material and product data for life cycle assessment in the built environment. International organisations such as the UN and the World Bank also implement an open data policy with datasets about their projects being freely and openly accessible. The United Nations Office for Project Services provides project-level datasets through the https://data.unops.org/ platform, and some of these datasets relate to construction and infrastructure projects. The United Nations Environment Programme, through the GEMS/Water Data Centre, provides a global freshwater quality database that is accessible at: https://gemstat.org/. Researchers from the civil engineering discipline apply such datasets to hydrological modelling and water resources planning. The World Bank also provides a similar service through their open data portal at: https://data.worldbank.org/, with datasets about issues such as sustainable development goals, climate change, infrastructure projects, and the like, which can be used for secondary research in the built environment research.

Social media data

The increased use of social media has resulted in large quantities of data that could be used to discover new social phenomena related to the built environment. This development provides the opportunity to extend traditional construction research into modern, social media-based big data analysis (Tang *et al.*, 2017). Through micro-blogging services such as Twitter, social scientists can mine and analyse data to gain unique insights into changes in social phenomena over time (Thelwall, 2014; Felt, 2016). Within the built environment context, construction organisations are using social media increasingly to communicate with stakeholders about what they do and how they do it; communities are using social media to share opinions during natural disaster situations, which can inform emergency response from a built environment perspective; construction workers are using social media to express their feelings about work; and most mega-construction projects have social media handles that are used to engage with stakeholders about ongoing construction. This provides vast opportunities for the use of pre-existing social media data for secondary research of topics such as stakeholder management and disaster risk management. The data can be processed using webometric analysis software such as Mozdeh (see http://mozdeh.wlv.ac.uk/), which can be used for time series graphing and analysis of a Twitter *corpus*, with the functionality to identify 1,000 individual words automatically within the *corpus* that exhibit the biggest increase in frequency over time (Thelwall, 2014). A sentiment analysis programme, such as SentiStrength, can be used to mine opinion to ascertain the sentiments conveyed in texts by using features such as the words used and the presence of emoticons (Thelwall, 2014). Other software packages such as NodeXL, Gephi, DMI Issue-Crawler can be used to perform analyses of social networks using data from social media (Felt, 2016). It should be noted that attempts to manipulate social media opinion for political goals (Thelwall, 2014) poses a challenge to this practice, with the potential for public opinions about the phenomena of interest being skewed.

Legal databases

There are databases that provide pre-existing data for legal research, which can be exploited by researchers in construction and engineering law. Legal research can involve a combination of data from both primary and secondary sources. Primary data in legal research take the form of primary sources of law such as legislation, and case law although these are already pre-existing. Secondary sources of law such as law reports, legal commentaries, and other literature about law are considered to be secondary data, which can be used when analysing legal principles and concepts. Westlaw, Lexis Library, Hein Online and British and Irish Legal Information Institute are examples of legal databases. Westlaw, for example provides an easily searchable source of case law, legislation, news, legal journals, and legal commentary, thus providing a source of both primary and secondary data for legal research.

Quality of secondary data

There are issues concerning the quality of secondary data that should be considered carefully before using the data for secondary research. To quote Smith (2008, p. 329), "as with all data, numeric or otherwise, an awareness of its limitations and a 'healthy scepticism' about its technical and conceptual basis is essential". Secondary data must be evaluated to ensure that it meets quality requirements, such as being accurate, reliable, valid, appropriate, and timely (Rabianski, 2003), just as with primary data. In terms of appropriateness, steps will have to be taken to ensure that the data measures what it was intended to measure. Where the pre-existing data has originated from primary research, checks should be made to ensure that the proper procedures were followed in collecting, organising, and analysing the original data that are now available as secondary data. This will require an understanding of contextual issues such as the response rate, inclusion, and exclusion criteria, the sampling pool, methods, and other relevant aspects of the original study so that these can be factored into the secondary study (Doolan *et al.*, 2017). Secondary data must also be checked to ensure that it is accurate for the topic being studied. This includes problems of data-fit. Secondary data may not be fully appropriate for the research purpose (e.g. missing information or information that is available in different type or format than what is needed for the research); thus, there is a need to consider if primary data can be used to meet the new aims of a secondary study.

There are also methodological challenges (e.g. lack of social context of research data). For more details, see for example, Feldman and Shaw (2019) and Ruggiano and Perry (2019).

There should also be awareness of the danger of bias being introduced into some pre-existing datasets (e.g. data from social media, government data), because of potential manipulation of data for political purposes in an era of post-truth and fake news. To ensure the reliability of secondary data, it should be reproducible, ensuring that if the systematic processes used to obtain and analyse the data are repeated, the same or similar results would be achieved. With research that relies on API-generated, social media datasets such as data from Twitter, there are limitations in this regard owing to the unknown logic of the algorithm that is used to produce the data (Felt, 2016). Also, it might not be possible to replicate Twitter datasets fully using API tools, as this is constantly changing, making it difficult to replicate and verify the findings independently (Felt, 2016). In terms of timeliness, data must be checked to ensure that they reflect the time period that governs the analysis (Rabianski, 2003), avoiding the use of outdated data in exploring time-sensitive research questions. As a general guide, Johnston (2014) suggested that to apply secondary data for research successfully, a systematic process is required, in which the challenges of the existing data are acknowledged and the distinct characteristics of the data are addressed within the analysis.

There are ethical issues that also need to be observed when using secondary data. These include adequate citation of the data source. It is important to cite the source of the data using the Digital Object Identifier, if this is available. Further ethical issues that must be considered before using secondary data for research have been covered in detail in Chapter 3 of this book.

Conclusion

In this chapter, secondary data have been discussed and contrasted with primary data, before providing an overview of potential secondary data sources that could be used for secondary research in the built environment. These sources comprise academic databases, government databases, databases of inter-governmental agencies, organisational databases, data from social media, and legal databases. All of these databases contain datasets that can be used in conducting secondary research in the built environment. However, these secondary sources of data do have some limitations, just as with primary data. To ensure that data from secondary sources are suitable for conducting secondary research, steps must be taken to ensure that quality requirements, such as accuracy, reliability, validity, appropriateness, and timeliness of the data, have been addressed. It is always important to demonstrate an awareness of the limitations of the secondary data, whilst ensuring that steps are taken to factor these limitations into the secondary research process.

References

Bishop, L. and Kuula-Luumi, A. (2017). Revisiting qualitative data reuse: A decade on. *SAGE Open*, pp. 1–15. doi:10.1177/2158244016685136.

Church, R.M. (2001). The effective use of secondary data. *Learning and Motivation*, 33, pp. 32–45. doi:10.1006/lmot.2001.1098.

Doolan, D.M., Winters, J. and Nouredini, S. (2017). Answering research questions using an existing data set. *Medical Research Archives*, 5(9), pp. 1–14.

Feldman, S. and Shaw, L. (2019). The epistemological and ethical challenges of archiving and sharing qualitative data. *American Behavioral Scientist*, 63(6), pp. 699–721.

Felt, M. (2016). Social media and the social sciences: How researchers employ Big Data analytics. *Big Data & Society*. doi:10.1177/2053951716645828.

Gebre, E.H. and Morales, E. (2020). How accessible is open data? Analysis of context-related information and users' comments in open datasets. *Information and Learning Sciences*, 121(1/2), pp. 19–36.

Greenhoot, A.F. and Dowsett, C.J. (2012). Secondary data analysis: An important tool for addressing developmental questions. *Journal of Cognition and Development*, 13, pp. 1, 2–18.

Heaton, J. (2008). Secondary analysis of qualitative data: An overview. *Historical Social Research*, 33(3), pp. 33–45.

Irwin, S. and Winterton, M. (2012). Qualitative secondary analysis: A guide to practice timescapes methods. *Guides Series*, Guide No. 19. Available at: https://timescapes-archive.leeds.ac.uk/wp-content/uploads/sites/47/2020/07/timescapes-irwin-secondary-analysis.pdf

Johnston, M.P. (2014). Secondary data analysis: A method of which the time has come. *Qualitative and Quantitative Methods in Libraries*, 3, pp. 619–626.

Largan, C. and Morris, T. (2019). *Qualitative Secondary Research: A Step-by-Step-Guide*, Sage, Thousand Oaks, CA.

Rabianski, J.S. (2003). Primary and secondary data: Concepts, concerns, errors and issues. *The Appraisal Journal*, 71(1), pp. 43–55.

Ruggiano, N. and Perry, T. (2019). Conducting secondary analysis of qualitative data: Should we, can we, and how? *Qualitative Social Work*, 18(1), pp. 81–97.

Sherif, V. (2018). Evaluating pre-existing qualitative research data for secondary analysis. *Forum Qualitative Sozialforschung / Forum: Qualitative Social Research*, 19(2), Art. 7. doi:10.17169/fqs-19.2.2821.

Smith, E. (2008). Pitfalls and promises: The use of secondary data analysis in educational research. *British Journal of Education Studies*, 56(3), pp. 323–339. doi:10.1111/j.1467–8527.2008.00405.x.

Tang, L., Zhang, Y., Dai, F., Yoon, Y., Song, Y. and Sharma, R.S. (2017). Social media data analytics for the U.S. Construction Industry: Preliminary study on Twitter. *Journal Management Engineering*, 33(6). doi:10.1061/(ASCE)ME.1943-5479.0000554.

Thelwall, M. (2014). Sentiment analysis and time series with Twitter. In: K. Weller, A. Bruns, J. Burgess., M. Mahrt and C. Puschmann. (eds.), *Twitter and Society*, Peter Lang, New York, pp. 83–95.

Tight, M. (2019). *Documentary Research in the Social Sciences*, 1st edition, Sage, Thousand Oaks, CA.

Zuiderwijk, A., Janssen, M. and Susha, I. (2016). Improving the speed and ease of open data use through metadata, interaction mechanisms, and quality indicators. *Journal of Organizational Computing and Electronic Commerce*, 26(1/2), pp. 116–146.

3 Ethical considerations in the use of secondary data for built environment research

Abid Hasan

Introduction to ethics in research

The root word for ethics is the Greek *ethos*, which means habit, character, or disposition (Aguinis and Henle, 2002). Liamputtong (2011, p. 25) defines ethics in research as "a set of moral principles which aims to prevent the research participants from being harmed by the researcher and the research process". Ethics in research provides guidelines for conducting, reviewing, and evaluating research, and establishes enforcement mechanisms to ensure ethical research. Although following the ethical norms and expectations in research can be a time-consuming and challenging process, complying with ethical principles and guidelines at each stage of research ensures the safety, rights, dignity, and well-being of both the participants and the researcher.

The scope of ethical consideration in research has widened across various disciplines during the past decades. Many academic institutions and research funding bodies now have dedicated ethics committees or institutional review boards to ensure that both the proposed research and the researchers adhere to the requirements of the national and international ethical guidelines and framework. However, the ethical requirements vary considerably, both across the disciplines and within a discipline, depending on the nature of the inquiry. For instance, disciplines such as clinical research need meticulous adherence to ethical principles because of its use of human and animal subjects and the invasive nature of the inquiry. By comparison, research in the built environment, though sometimes involving human participants, requires less stringent ethical considerations because of the non-invasive nature of the investigation. Built environment researchers normally do not collect identifiable human material, such as human tissue or medical data, or conduct human or animal trials. However, they might collect identifiable information from the participants in the form of age, experience, designation, and place of work. Moreover, the researchers often approach human participants directly or indirectly to collect data. Therefore, it is necessary for the built environment research community to follow ethical principles that might apply to research involving human subjects.

According to the principle of non-maleficence, researchers must ensure the physical, emotional, and social well-being of research participants (Ramcharan, 2010). Peterson (2013, p. 8) describes the researcher-participant relationship as sacrosanct and similar to a lawyer-client or doctor-patient relationship, in that the privacy of study participants must be respected and identities must be protected by keeping responses anonymous and confidential in all stages of research unless extenuating circumstances exist. Therefore, researchers must ensure that the data collected are strictly confidential and anonymous, and participants are informed about the steps that will be taken by the researcher to maintain confidentiality and anonymity (Holloway and Wheeler, 2010). Moreover, explicit and informed consent from the participants must be obtained before they participate in the research. Using human data in secondary data research without following a rigorous ethical process can have serious repercussions for built environment researchers, participants, and other stakeholders. There is always a risk that individuals or cohorts with distinctive characteristics can be identified even within existing, large-scale datasets.

Stewart *et al.* (2017) found minimum concern for human research ethics in construction management research owing to lack of evidence of concern for human research ethics together with the absence of any publisher demands or institutional standards. However, it does not mean that built environment researchers do not follow ethical requirements at different stages of the research. In fact, the publishing requirements in the built environment are not very stringent and vary significantly amongst journals, and the results are sometimes published without explicit discussions of the ethical requirements and the steps taken to ensure that the research adheres to ethical guidelines. Similarly, for research using primary data collection methods, such as questionnaire surveys, focus groups or interviews, many institutions have a straightforward and well-defined process to ensure that research and researchers meet ethical standards. However, ethical requirements can be ambiguous in the context of research based on secondary data. In the absence of well-established procedure and guidelines, the need to exercise ethical controls can be overlooked sometimes in secondary research.

Some of the important ethical considerations that apply to built environment research based on secondary data will be highlighted in this chapter. The interest in the use of secondary data for research in the built environment has grown considerably in recent times. Understanding what constitutes ethical research, in the context of using secondary data and the ethical process, is an important step to incorporating ethical considerations in all phases of secondary data research, from the research design to analysis and presentation. Furthermore, adherence to ethical principles will improve transparency and accountability in research using secondary data. This chapter will also include an outline of ethical practices that built environment researchers could incorporate at different stages of secondary data research.

Secondary data research in the built environment

Existing data, collected by someone else for another primary purpose, are identified as secondary data. Therefore, the researchers in secondary data research do not collect the data empirically (Johnston, 2017). An enormous amount of data is now generated, collected, compiled, shared, and archived as a result of the growth in data management infrastructure and capabilities. Research utilising secondary data

provides many opportunities for furthering built environment research through replication, triangulation, re-analysis, and re-interpretation of existing research. Built
environment researchers can test new ideas, theories, frameworks, and models of research design using secondary data. Also, the used or unused data can provide new
insights or different perspectives in secondary research (Heaton, 2008; Johnston,
2017). The utilisation of existing data for research is becoming a more viable and
prevalent option for researchers who might have limited time and resources. The
use of secondary data can accelerate the pace of research owing to the availability
of data (Doolan and Froelicher, 2009). For instance, primary data collection is one
of the challenging time- and resource-demanding steps in a typical built environment research project. Surveys in built environment research often suffer from low
response rates, possibly because of the busy work schedules of potential respondents.
Bing *et al.* (2005) observed that response rates as low as 10–12% were not untypical
in construction management research. In these circumstances, secondary data can
be used either as a substitute or complement for primary data.

The secondary data can take different forms in the context of the built environment. In addition to survey, interview or case study data collected by other researchers, data available in the form of documentary records, case notes, archives,
and other data can be used by built environment researchers for secondary data
analysis. For instance, reports and data published by governmental and regulatory bodies contain both micro- and macro-level data and social and industry
statistics. There are also many quasi-governmental and other official bodies that
produce large amounts of data. For instance, published legal judgements can be
used to provide rich insights into disputes and dispute resolution mechanisms in
construction projects. Similarly, annual reports and newspaper articles can be
used for content analysis. Construction organisations can be one of the most important sources of secondary data, through publicly available annual reports and
financial data. Moreover, the extensive use of online communication and collaboration platforms by individuals and organisations generates a tremendous amount
of archival information that can be used by researchers to provide insights into
a variety of phenomena such as project and organisational communication and
stakeholder management. Table 3.1 shows some of the examples of built environment research based on the analysis of secondary data.

In some cases, the detailed datasets such as industry and economic data are
available to researchers to perform secondary analysis, whereas in other cases,
researchers can only access aggregate data, or the findings based on the analysis
of primary data. For instance, the raw data held by the regulatory bodies, such
as health and safety councils, generally are not available publicly. Consequently,
the insights into the underlying data can be somewhat limited and might not
present meaningful opportunities for secondary analysis. Therefore, researchers
often need access to the underlying data rather than published, analysed results
to perform rigorous analysis. On most occasions, researchers can request access to
the primary data, and then it depends on how much information the data custo-
dians are willing to share for secondary analysis. The identifying information that
was collected initially is sometimes removed before sharing the original dataset

Table 3.1 Examples of built environment research based on secondary data

Reference	Purpose of the study	Secondary data used in the study	Source(s) of data
Azhar *et al.* (2019)	To examine the current usage of social media within the construction industry using both primary and secondary data	Construction companies' social media activity on Facebook, LinkedIn, and Twitter	Facebook, Twitter, and LinkedIn
Hamie *et al.* (2018)	To offer a holistic model for articulating the contract clause addressing the respective priorities of the contract documents	Standard sets of contract conditions, coupled with a review of related practices found in the literature	The American-based sets by the American Institute of Architects, Engineering Joint Contract Documents Committee, and ConsensusDOCS; the British-based forms of the Joint Contract Tribunal and New Engineering Contract; the internationally used standard contract conditions by the International Federation of Consulting Engineers
Hasan *et al.* (2018)	A systematic review of studies on identifying factors affecting construction productivity published during the last three decades	Published articles	Journals, conference proceedings, dissertation, and PhD theses
Hinze and Teizer (2011)	To examine vision-related construction fatalities and identify various circumstantial factors	Fatality accidents data from a data pool of 13,511 Occupational Safety and Health Administration (OSHA)-investigated cases	OSHA – United States
Faghih and Kashani (2018)	To present a vector error correction model to forecast the short- and long-term prices of construction materials	Cement producer price index; consumer price index; hourly earnings of construction labour; industrial producer price index; number of employees in construction; and so on.	US Bureau of Labour Statistics; US Bureau of Economic Analysis; US Bureau of Labour Statistics; US Census Bureau; and so on

(Continued)

Reference	Purpose of the study	Secondary data used in the study	Source(s) of data
Sing et al. (2015)	To provide a study of the annual financial value of construction work in the private residential market using a vector auto-regression model	Best lending rate; property price index; vacancy rate; and so on	The Hong Kong Monetary Authority; the Rating and Valuation Department of the Hong Kong Government; and so on
Mason (2017)	Scope out the conditions for successful adoption intelligent contracts in the construction industry and the resulting benefits	Responses from individual contributors to the debate	Online forums where the identities of the contributors are unknown to the viewer
Moza and Paul (2017)	To evaluate the effectiveness of the current arbitration procedure in India, specifically in the public sector	Cases of arbitration	Central Public Works Department, India
Nguyen et al. (2019)	To explore indirect pathways pursued by the external stakeholders to affect construction projects and develop a framework of stakeholder-influencing pathways	Newspaper articles, press release, petition, blogs for four case studies on construction projects in Vietnam	Newspaper and archived data, such as press releases, petitions, blog postings, and official documents

Source: Original.

to protect the privacy of participants. Moreover, researchers might need to sign a contract to ensure that data will be re-used within appropriate ethical and legal boundaries for research purposes only.

Ethical considerations in secondary data research

If data is collected for one purpose, can it be used for other purposes? Is it acceptable to keep data forever? Will explicit consent be requested for the use of publicly available, secondary data?

These common ethical considerations in secondary data research will depend on the nature of the existing data to be used in the research. Not all types of research based on secondary data would be required to meet all ethical requirements. Therefore, it is essential to examine how ethical principles should be applied to a specific topic or question. For instance, secondary analysis of publicly available data on the energy performance of buildings, industry data, organisational annual performance reports, and public policies might not need to meet some of the ethical principles that apply to data collected directly from human participants. Nonetheless, the dataset in the public domain (e.g. personal social media accounts) should not always be considered to be free from restrictions and limitations on its use or re-use. Also, certain datasets available in the public domain include the terms and conditions of re-use or contract from the data supplier or source. The researchers in the built environment must adhere to working within limits agreed or covered by the licence.

For research based on secondary data involving human participants or human subjects, built environment researchers should apply the Belmont Principles. The three Belmont Principles, published in 1978 and widely consulted to establish ethical guidelines for human research, are: *autonomy* (respect for persons), *beneficence*, and *justice* (Levine, 1986; Rafi and Snyder, 2015). Stewart *et al.* (2017) also discussed a five-item framework that constitutes ethically sound research in construction management research involving human participants. The framework includes autonomy, beneficence, non-maleficence, confidentiality, and integrity. Figure 3.1 shows the main ethical considerations relevant to built environment researchers working on secondary data either collected initially from human participants or data available in the public domain.

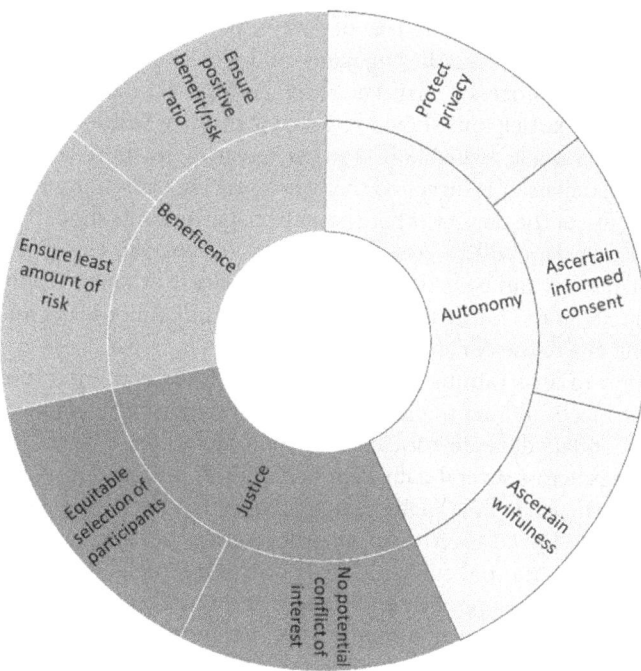

Figure 3.1 Ethical considerations for built environment research on secondary data.
Source: Original.

Autonomy

The secondary analysis of data poses ethical concerns about informed consent if the consent applies to the original study only. If datasets are reported with a statement about the re-use of the data, then researchers have an ethical duty to comply with those requirements. For most databases, researchers will have to search for the permission statement on the use and re-use of the data (for instance, http://www.nationalarchives.gov.uk/doc/open-government-licence/version/3/ for the UK Government database). The secondary data analysis must abide by the consent conditions of the original study (Heaton, 2008). However, in instances where information about use or re-use of the data is unavailable, the ambiguity about what kind of consent is required or whose permission is required to conduct research on secondary data can create ethical issues for built environment researchers. If the original participants were not well informed about the re-use of data in future research, including participants and their data in a secondary research study raises ethical concerns about informed consent. Moreover, in some cases, obtaining informed consent could be a time-consuming and costly exercise. It could also be practically impossible to contact data subjects again to obtain consent for secondary analysis of their data (Bishop 2017).

In the last few years, the importance of sharing raw data has been gaining momentum in the built environment discipline. Many reputed journals now include a data availability statement to encourage the sharing of data among researchers. Therefore, it is recommended that when collecting primary data, researchers plan upfront for archiving and future uses of data by paying attention to study protocols, informed consent, data-sharing plans, and preserving a detailed account of the data collection process (Corti *et al.*, 2014; Antes *et al.*, 2018). During the data collection stage, participants should be asked if they are willing for their data to be archived and made available for further research. In the UK Data Archive Guide for researchers, it is suggested that when seeking consent from participants for future re-use of the data by other researchers, participants should be informed of how research data will be stored, preserved, and used in the long term, and how confidentiality can be protected when necessary. Without providing specific information about the future use of data, participants cannot make informed decisions about the re-use of data by other researchers.

In addition to ascertaining wilfulness and informed consent, protecting the privacy of subjects and maintaining confidentiality during all stages of research based on secondary data are critical ethical considerations. The ability to match or link records across several data sources has implications for privacy and data security (Salerno *et al.*, 2017). Different attributes that are non-identifiable independently, when linked together using other sources of data, can make participants identifiable (Yiannakoulias, 2011). For instance, social media posts quoted verbatim can easily be searched and thereby can reveal the identity of the subjects without their informed consent. The disclosure of sensitive information could result in financial, criminal, or civil liability, or damage the subjects' personal and professional lives. While the use of pseudonyms or replacements can protect a subject's identity, the data-sharing policy of the social media platform (e.g. Twitter developer agreement/policy) typically prohibits modifying content. Swirsky *et al.* (2014) discuss the ethical issues pertaining to the use of social media for research purposes because many users might not fully understand the privacy implications. Therefore, the disclosure of identifiable data in secondary research that can be linked to other information to disclose private and sensitive information can cause serious harm to both subjects and built environment researchers.

Beneficence

According to the principle of beneficence in research, the maximum benefit should be delivered to participants and society at large, while minimising any possible harm (both physical and psychological). While primary data collection methods, such as surveys, interviews, and observations, might cause discomfort and frustration for participants, gathering secondary data does not involve direct or indirect interactions with human participants. For instance, different types of secondary data (e.g. social media data, open industry data, and policy documents) are publicly available for research without involving any direct interaction with the subjects. The risks to the participants, during secondary data analysis and dissemination of findings, can be further minimised by ensuring privacy, confidentiality, and anonymity of the data.

The principle of beneficence includes the recommendation to share and re-use data to their full extent to examine multiple research questions to maximise the benefits of an individual subject's participation and to extract the most knowledge possible (Brakewood and Poldrack, 2013). While the secondary analysis of data follows the fundamental principle of beneficence because the intention is to use or re-use the data to provide new insights, the researchers and institutional ethics review committee must ensure that the secondary data research makes a meaningful contribution to the field of study. The research problem should justify the need for the analysis of secondary data to meet the beneficence criterion of positive benefit to risk ratio.

Justice

Justice, in research, refers to a balance of benefit and burden. According to the principle of justice, the selection of data or participants should ensure that individuals or groups are used fairly in research (Brakewood and Poldrack, 2013). Ideally, each member of the population should have an equal probability of being chosen for the findings to be a reliable and true representation of the population (Preston, 2009), and sampling should be done in such a way that the researcher does not influence the selection of the respondents (Saris and Gallhofer, 2014). However, owing to lack of a common public database of contact details for built environment professionals, resource constraints and other practical considerations, probability, or simple random sampling often is not feasible in built environment research. Consequently, secondary analysis of existing empirical datasets often is not based on data collected from a random sample, which might affect the generalisability and scientific validity of findings. However, pooling secondary data from different subjects, sources, and populations can increase the generalisability of findings and conform to the principle of justice in secondary data research (Brakewood and Poldrack, 2013). For instance, researchers can access annual financial data of all construction companies listed on the national stock exchange or obtain information on all recorded fatalities on construction sites, over a period, from the workplace health and safety department to minimise the selection bias. Therefore, in research based on secondary data obtained from sources such as government datasets, social media platforms, and industry reports, it is easier to comply with the principle of justice through access to Big Data and a large sample.

Ethical obligations during different stages of secondary data research

Where applicable, ethical considerations should apply to the whole research process from first thoughts to dissemination through publications. Figure 3.2 illustrates the main stages of a secondary research design to be considered by built environment researchers to meet the ethical obligations in research based on secondary data.

Conceptual phase

Just as in primary research, the quality of the research is an essential, ethical consideration in research based on secondary data. If the study is poorly designed or if it has no value, the outcomes of the investigation will be of no benefit to society (George, 2016). For instance, literature review papers limited to summarising or re-stating the findings of previous research, without providing critical insights into the current state of the research and future directions, offer limited value to the readers. The secondary data researchers have obligations to produce and disseminate high-quality research by adopting a rigorous, respectful, and robust research process (Camfield, 2019). Without a distinct contribution to the body of knowledge and novelty in research, it is also difficult to publish the findings

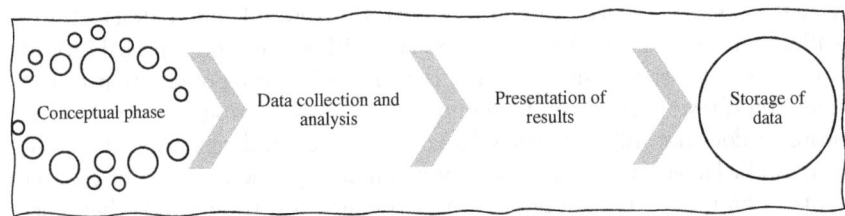

Figure 3.2 Main stages in secondary data research from the ethical perspective.
Source: Original.

of research in a good-quality, peer-reviewed journal. Therefore, secondary data research in the built environment must be scientifically sound, rigorous, and designed to produce a benefit for the field of study.

Data collection and data analysis

Before using secondary data in research, researchers must ensure the appropriateness and quality of data in the context of their research because they cannot exercise any control over the generation of data (Harris, 2001). The existing datasets might not have been collected in the geographic region of interest or in the years of interest or about the specific population of interest (Boslaugh, 2007; Doolan and Froelicher, 2009). Similarly, qualitative data is context-specific and relationship-dependent (Broom *et al.*, 2009; Yardley *et al.*, 2014). Irwin (2013) discusses ethical concerns related to the secondary analysis of qualitative data collected from in-depth interviews, ethnographic inquiries, and fieldwork, where data are generated through personal interactions between researchers and participants. The secondary analysis of such data and the validity of the interpretations without immersion in the data and awareness of the context surrounding the data collection have been questioned by many researchers (Irwin, 2013).

Built environment researchers can follow best practices for collecting and sharing data to address these concerns to a large extent. For example, it is essential to understand the context of original research with the help of field notes and diaries, recorded observations, technical reports, publications, and reflections providing a detailed description of the data collection process so that secondary researchers can use the data appropriately (Corti *et al.*, 2014; Camfield, 2019). Knowledge of the primary method used by the original researchers could address validity and reliability issues in secondary research and help in performing new analyses on existing datasets (Boslaugh, 2007; Johnston, 2017).

Dissemination of findings

While reporting the findings of a secondary data analysis, researchers should include information on the original research design, such as a detailed description of the collection of original data, its sources and processing to show adherence to ethical standards. The ownership of the original data must be acknowledged. The specific details about sampling protocol and data collection procedures, response rate, sampling bias, missing data, and the time frame of data collection should be reported (Boslaugh, 2007). Researchers have a responsibility to show rigour and transparency by providing a detailed account of how the research was conducted and how analysis and interpretation were carried out (and by whom), and also the evidence to support the claims made, based on secondary data analysis (Manski, 2011). Researchers must also remember that linking data from different sources might reveal the identity of the individuals or groups involved in the study. It is crucial to use acceptable and effective methods of ensuring data anonymity while reporting the findings of the research. For example, instead of directly quoting identifiable, individual, social media content, the researcher can present the analysis results in aggregate forms such as word clouds, sentiment analysis, and thematic analysis to preserve anonymity and confidentiality (Williams *et al.*, 2017).

Data storage

Researchers should remove all identifying information prior to storing or sharing the data for re-use. It should be indicated whether the material may be used for research purposes and what the procedure is for gaining consent (Grinyer, 2009). The data should be stored in the de-identified form to meet the ethical guidelines for privacy, confidentiality, and anonymity of participants. Researchers must adopt rigorous data security protections to maintain confidentiality and protect the privacy of participants. Moreover, country-specific ethical and legal requirements concerning privacy, data storage, and data protection must be considered in internet-mediated built environment research using Big Data and social media.

Conclusions

Following recent advances in data management and digital technologies, it is expected that the use of secondary data for answering potential research questions will increase in built environment research. The secondary analysis of data can offer both validation and replication of original studies as well as provide new research opportunities to built environment researchers. Consequently, it is necessary that the tenets of the ethical conduct of research are examined and updated for research based on secondary data. Owing to the variety of type and sources of secondary data, it is difficult to arrive at standard codes of ethics. The extent of the application of ethical principles to secondary research would depend on the type of secondary data to be used in the research. The research protocol for secondary data research should address the questions regarding how the existing data will be collected, analysed, anonymised, published, stored, and secured. Specifically, built environment researchers must consider ethical matters concerning anonymity and confidentiality of subjects.

Researchers have an ethical duty to comply with: the terms and conditions specified by the data producer, depositor, and the data repositories for the use and re-use of secondary data; established institutional ethical research policies and guidelines; and fundamental ethical principles discussed in this chapter. Repositories, publication outlets, and institutions should support secondary data research in the built environment by publishing written guidance and policies to inform researchers on legal, institutional, and ethical requirements concerning sharing, archiving, and re-using data for research. In addition, experienced researchers and supervisors could assist students and novice researchers, who often are not equipped with skills or sensitivity to address ethical issues related to the use of secondary data. It is recommended that all projects based on secondary data analysis undergo research committee review, which would then determine the level of compliance with scrutiny and research ethics. It would be wrong to presume that ethical concerns do not apply to built environment research based on secondary data.

References

Aguinis, H. and Henle, C. (2002). Ethics in research. In: S.G. Rogelberg (ed.), *Handbook of Research Methods in Industrial and Organizational Psychology*, Blackwell Publishing, Malden, MA, pp. 34–56.

Antes, A., Walsh, H., Strait, M., Hudson-Vitale, C. and DuBois, J. (2018). Examining data repository guidelines for qualitative data sharing. *Journal of Empirical Research on Human Research Ethics*, 13(1), pp. 61–73.

Azhar, S., Riaz, Z. and Robinson, D. (2019). Integration of social media in day-to-day operations of construction firms. *Journal of Management in Engineering*, 35(1), p. 06018003.

Bing, L., Akintoye, A., Edwards, P. and Hardcastle, C. (2005). The allocation of risk in PPP/PFI construction projects in the UK. *International Journal of Project Management*, 23(1), pp. 25–35.

Bishop, L. (2017). Big data and data sharing: Ethical issues. UK Data Service, UK Data Archive. Available at: https://ukdataservice.ac.uk/media/604711/big-data-and-data-sharing_ethical-issues.pdf.

Boslaugh, S. (2007). *Secondary Analysis for Public Health: A Practical Guide*, Cambridge University Press, New York. doi:10.1017/CBO9780511618802.

Brakewood, B. and Poldrack, R. (2013). The ethics of secondary data analysis: Considering the application of Belmont principles to the sharing of neuroimaging data. *Neuroimage*, 82, pp. 671–676.

Broom, A., Cheshire, L. and Emmison, M. (2009). Qualitative researchers' understandings of their practice and the implications for data archiving and sharing. *Sociology*, 43(6), pp. 1163–1180.

Camfield, L. (2019). Rigor and ethics in the world of big-team qualitative data: Experiences from research in international development. *American Behavioral Scientist*, 63(5), pp. 604–621.

Corti, L., Van den Eynden, V., Bishop, L. and Woollard, M. (2014). *Managing and Sharing Research Data: A Guide to Good Practice*, Sage, London.

Doolan, D. and Froelicher, E. (2009). Using an existing data set to answer new research questions: A methodological review. *Research and Theory for Nursing Practice: An International Journal*, 23(3), pp. 203–215. doi:10.1891/1541–6577.23.3.203.

Faghih, S. and Kashani, H. (2018). Forecasting construction material prices using vector error correction model. *Journal of Construction Engineering and Management*, 144(8), p. 04018075.

George, A. (2016). Research ethics. *Medicine*, 44(10), pp. 615–618.

Grinyer, A. (2009). The ethics of the secondary analysis and further use of qualitative data. *Social Research Update*, 56(4), pp. 1–4.

Hamie, J. and Abdul-Malak, M. (2018). Model language for specifying the construction contract's order-of-precedence clause. *Journal of Legal Affairs and Dispute Resolution in Engineering and Construction*, 10(3), p. 04518011.

Harris, H. (2001). Content analysis of secondary data: A study of courage in managerial decision making. *Journal of Business Ethics*, 34(3–4), pp. 191–208.

Hasan, A., Baroudi, B., Elmualim, A. and Rameezdeen, R. (2018). Factors affecting construction productivity: A 30 year systematic review. *Engineering, Construction and Architectural Management*, 25(7), pp. 916–937.

Heaton, J. (2008). Secondary analysis of qualitative data: An overview. *Historical Social Research*, 33(3), pp. 33–45.

Hinze, J. and Teizer, J. (2011). Visibility-related fatalities related to construction equipment. *Safety Science*, 49(5), pp. 709–718.

Holloway, I. and Wheeler, S. (2010). *Qualitative Research in Nursing and Healthcare*, 3rd edition, Wiley-Blackwell, Oxford.

Irwin, S. (2013). Qualitative secondary data analysis: Ethics, epistemology and context. *Progress in Development Studies*, 13(4), pp. 295–306.

Johnston, M. (2017). Secondary data analysis: A method of which the time has come. *Qualitative and Quantitative Methods in Libraries*, 3(3), pp. 619–626.

Levine, R. (1986). *Ethics and Regulation of Clinical Research*, 2nd edition, Urban and Schwarzenberg, Baltimore, MD.

Liamputtong, P. (2011). *Focus Group Methodology: Principles and Practice*, Sage, London.

Manski, C. (2011). Policy analysis with incredible certitude. *Economic Journal*, 121, pp. F261–F289.

Mason, J. (2017). Intelligent contracts and the construction industry. *Journal of Legal Affairs and Dispute Resolution in Engineering and Construction*, 9(3), p. 04517012.

Moza, A. and Paul, V. (2017). Review of the effectiveness of arbitration. *Journal of Legal Affairs and Dispute Resolution in Engineering and Construction*, 9(1), p. 03716002.

Nguyen, T., Chileshe, N., Rameezdeen, R. and Wood, A. (2019). Stakeholder influence pathways in construction projects: Multi-case study. *Journal of Construction Engineering and Management*, 145(9), p. 05019011.

Peterson, R. (2013). *Constructing Effective Questionnaires*, Sage, Thousand Oaks, CA.

Preston, V. (2009). Questionnaire survey. In: R. Kitchin and N. Thrift (eds.), *International Encyclopedia of Human Geography*, Elsevier, Oxford, pp. 46–52.

Rafi, N. and Snyder, B. (2015). Ethics in research: How to collect data ethically. In: M.P. Wilson, K.Z. Guluma and S.R. Hayden (eds.), *Doing Research in Emergency and Acute Care: Making Order Out of Chaos*, John Wiley & Sons, Ltd., Chichester, pp. 45–51.

Ramcharan, P. (2010). What is ethical research? In: P. Liamputtong (ed.), *Research Methods in Health: Foundations for Evidence-Based Practice*, Oxford University Press, Melbourne, pp. 27–41.

Salerno, J., Knoppers, B., Lee, L., Hlaing, W. and Goodman, K. (2017). Ethics, big data and computing in epidemiology and public health. *Annals of Epidemiology*, 27(5), pp. 297–301.

Saris, W. and Gallhofer, I. (2014). *Design, Evaluation, and Analysis of Questionnaires for Survey Research*, 2nd edition, John Wiley & Sons, Hoboken, NJ.

Sing, M., Edwards, D., Liu, H. and Love, P. (2015). Forecasting private-sector construction works: VAR model using economic indicators. *Journal of Construction Engineering and Management*, 141(11), p. 04015037.

Stewart, I., Fenn, P. and Aminian, E. (2017). Human research ethics – Is construction management research concerned? *Construction Management and Economics*, 35(11–12), pp. 665–675.

Swirsky, E., Hoop, J. and Labott, S. (2014). Using social media in research: New ethics for a new meme? *The American Journal of Bioethics*, 14(10), pp. 60–61.

Williams, M.L., Burnap, P. and Sloan, L. (2017). Towards an ethical framework for publishing Twitter data in social research: Taking into account users' views, online context and algorithmic estimation. *Sociology*, 51(6), pp. 1149–1168.

Yardley, S., Watts, K., Pearson, J. and Richardson, J. (2014). Ethical issues in the re-use of qualitative data. *Qualitative Health Research*, 24(1), pp. 102–113.

Yiannakoulias, N. (2011). Understanding identifiability in secondary health data. *Canadian Journal of Public Health*, 102(4), pp. 291–293.

4 Qualitative secondary analysis (QSA) as a research methodology

Victoria Sherif

Introduction

Qualitative Secondary Analysis (QSA) is a powerful methodological and analytical approach that is centred on the re-analysis of previously collected (existing) data. QSA is often referred to as "any research activity in which a researcher uses data for purposes not defined or predicted in the original study design" (Yardley *et al.*, 2014, p. 102). In the social and educational context of built environment, QSA enables further explanation, re-interpretation, or validation of existing findings and thus the development of knowledge that otherwise would have not been generated without secondary analysis (Carmichael, 2017; Irwin and Winterton, 2011; Sales, Lichtenwalter and Fevola, 2006). QSA encompasses the exploration of educational and social environments, thoughts, underlying reasons, motivations, and results of people's activities (Heaton, 2004). As with any other method of inquiry, the aim of QSA is to contribute to scientific knowledge by offering alternative theoretical or conceptual perspectives on previously collected data (Johnston, 2014). Using QSA also has the potential of enabling the researcher to consider important methodological questions without the limitations encountered in the original research (Mitchell, 2015). With an emphasis on lived educational and social experiences, QSA is fundamentally well suited to broadening the meanings people place on their lives and to identifying areas for further inquiry.

The main objectives and outcomes of QSA as well as considerations for selecting and working with existing qualitative data have been examined in this chapter. With a six decade-long history, the epistemological and methodological objectives and outcomes of QSA are becoming more defined. The importance of a systematic and comprehensive evaluation of existing qualitative data used for the purpose of QSA is reiterated in this chapter. This includes guiding questions related to dataset fit, relevance, quality, trustworthiness, and timeliness.

Brief history of qualitative secondary analysis

QSA has played a critical role in the development of qualitative research inquiry. The emergence of QSA has been dated back to the early 1960s, as social and educational researchers recognised the untapped potential of existing data to fulfil research needs without the burden of initiating a new study (Glaser,

1963). The recent popularity of QSA has been attributed to the increasing ac-cessibility and number of archived qualitative data sources (Beck, 2019; Bishop and Kuula-Luumi, 2017; Hughes and Tarrant, 2019; Vartanian, 2011). The de-velopment and integration of QSA in social and educational sciences were also driven by the time-consuming nature, sample sizes, and monetary commitments of qualitative research. As an attempt to minimise the collection of new data and mitigate extensive financial and time commitments (Holland *et al.*, 2006; Irwin and Winterton, 2014; Smith, 2008; Thomson and McLeod, 2015; Thorne, 2013), QSA emerged as an opportunity to utilise available, archived, qualitative data sufficiently and to contribute meaningfully to empirical scholarship.

Popularisation of data archiving changed attitudes towards QSA, first, by sav-ing the time required to produce extensive data and, second, by providing access to qualitative data founded on diverse sample sizes. As qualitative data archiving increased in scale, the discussion about QSA expanded to include topics such as why and how to archive and use archived data, ensure technical and ethical transparency, and maintain participant confidentiality. Increased data archiving contributed to the development of approaches to existing data analysis and fos-tered building the capacity of QSA as a method in social and educational inquiry.

The advancement of archiving and the popularity of QSA inspired methodo-logical debates about: the possibilities and challenges of QSA (Beck, 2019; Hinds, Vogel and Clarke-Steffen, 1997; Irwin, 2013; Mauthner *et al.*, 1998); differences and boundaries between forms of primary and secondary qualitative analysis (Corti, 2012; Heaton, 2000, 2004; Moore, 2007); data sharing (Stamm, 2018); and crea-tive approaches of using multiple, existing datasets (Andrews, 2008; Neale, 2019). Following these epistemic and methodological shifts, QSA became more accepted, thus creating more opportunities for reflexivity and theoretical development of the method (Akcam, Guney and Cresswell, 2019; Irwin, Bornat and Winterton, 2012).

From an epistemological perspective

QSA requires continuous reflection on how to re-contextualise and re-purpose previously collected data for new research. Using existing qualitative data for sec-ondary analysis involves reorganising the data as independent constructs, identi-fying how the data were generated, and determining how the data will be re-used. Being systematic and subjective in nature, QSA also includes reflexivity to high-light and explain human experiences that emerged in a given social and/or edu-cational context with little to no investment in context re-creation. Reflection on previously explored topics from a new perspective can enrich the understanding of a subject as well as contribute to the generation of more meaningful, empirical, and/or methodological findings.

From a methodological perspective

QSA may encompass multiple stages of data preparation and evaluation (Heaton, 2008; Irwin, 2019; Irwin and Winterton, 2012; Sherif, 2018a). Since existing data are embedded in the contexts in which they are gathered (Janzen, 2012), the details present in the existing dataset can be used to inform the trustworthiness of available data and pre-determine the capacity of QSA to meet new research goals sufficiently (Corti and Blackhouse, 2005; Corti and Bishop, 2005). It is critical to explore available data rigorously so to understand how and why the existing dataset was generated, what it does and does not include, and how the dataset relates to new QSA inquiry. Therefore, extensive preparatory and evaluative work is required in order to address secondary research objectives adequately and meaningfully.

While preparing and evaluating existing qualitative data, it might be found that the purpose of the original research does not fit the purpose of the QSA study, thus limiting the scope of available data. Specific features of qualitative data used for QSA might also be unavailable or scarce and thus be a challenge to the reconstruction of the context of the original study. Furthermore, the data might be out of date and no longer appropriate for the objective(s) of the secondary research.

Objectives and outcomes of qualitative secondary analysis

In this section, the main objectives and outcomes of QSA that secondary analysts must consider when designing and conducting a QSA study are examined. Depending on the nature, quality, depth, and breadth of available qualitative data as well as researchers' interests, the objectives and resulting outcomes of QSA fall into three main categories (see Table 4.1). First, if the aim is to generate new

Table 4.1 QSA objectives and outcomes

Objective	Description	Outcome
Generation of new knowledge	Investigation of existing qualitative data through new research question, theoretical orientation, or by using/combining multiple available datasets on the same topic of interest to broaden the knowledge about the subject, conceptualise, or validate the original findings	Expanded understanding of the topic of interest or a concept that was not examined at the time of the original study, topic conceptualisation, validation of previously generated research findings, and/or their comparison
Design of new inquiry	Using existing qualitative data to conceptualise a new study and/or define supplemental data collection	New research questions; new or expanded dataset
Generation of methodological insights	Examination of existing qualitative data to advance QSA conceptually as a discrete method	Expanded and/or refined procedures of original data assessment, ethical considerations, archiving, and analysis

Source: Original.

knowledge, QSA can make it possible to re-analyse data through new research questions, theoretical or methodological approaches, and diverse datasets. Second, if the aim is to broaden the scope of research conducted previously, QSA can be used to help in the design of a new inquiry to build on the existing data. Third, if the aim is to expand methodological and conceptual underpinnings, QSA can be used to reveal new complexities of the method ethics, assessment of existing data, and procedures of data analysis and archiving. This section begins with a description of these objectives and outcomes. Examples of QSA studies conducted in the social and educational context of built environment are included to illustrate these objectives and outcomes.

One of the compelling reasons for undertaking QSA is to generate new knowledge or broaden understanding of a topic of interest. QSA can be effective when "there is a little information known about a phenomenon, the applicability of what is known has not been examined, or when there is a reason to doubt the accepted knowledge about a given phenomenon" (Kidd, Scharf and Veazie, 1996, p. 225). As a result, it can generally broaden and deepen knowledge by stimulating a comprehensive understanding of the nature of an examined issue. Secondary analysis of qualitative data can be used also to reveal additional contexts of educational and social encounters and collaborations as well as inform the situation of narratives around a particular topic.

Secondary analysis of qualitative data allows for both re-interpretations and the introduction of new research questions and/or a new theoretical framework. New perspectives can be applied which might not have been considered at the time of the original data collection. Furthermore, new research questions and theoretical orientations can be used to enunciate aspects of data that were dismissed or overlooked previously in the primary analysis. A new conceptual and theoretical approach to the analysis of existing qualitative data can be used to re-organise the data, thus creating a possibility for a QSA study.

An example of how secondary analysis of existing qualitative data led to the generation of new knowledge found in the secondary inquiry by Anderson (2019) to explore leadership actions and activities that contributed to the implementation of family engagement initiatives in an elementary school. The existing qualitative data included a survey, documents, and correspondence materials. The data were collected to inform the design and implementation of professional development related to family engagement, ensure accountability for implementation, and to build the capacity of staff and families to support student learning. Several leadership activities related to family engagement were also identified: communication with staff and families, formal assessment, professional development, and leader meetings.

Another example of a QSA study conducted by Gupta and Sherif (2021) illustrates how application from a new theoretical perspective increases potential to generate new knowledge. The original qualitative study by Gupta (2010) on the supports and challenges in collecting accurate and reliable data about educational outcomes for children to meet federal accountability requirements was guided by a policy-based approach and was focused on factors and processes to ensure collection of high-quality data about educational outcomes for children. Semi-structured interviews, using open-ended questions, were conducted with 42 co-ordinators responsible for implementing early childhood special education programmes. For the purpose of the QSA study, a statement of the implementation of an early childhood professional development system by leaders of early intervention services was viewed from a positive organisational scholarship perspective introduced in the study. There were several overarching themes that indicated how early intervention leaders transformed challenges into opportunities. The study involved meeting practitioners where they were, identifying leaders, and using consistent measurement procedures. Additional themes were focused on the importance of relationships and training during the transformative process of early interventions.

With the aim of generating new knowledge, QSA can be applied to multiple datasets to compare or validate findings generated in the original study as well as to gain new insights through the diversity and depth of available datasets (Heaton, 2008). Essentially using QSA makes it possible to carry out case-based analyses to build conceptual understanding of an explored phenomenon across different social and educational contexts (Irwin, 2019; Tarrant and Hughes, 2019). The newly generated knowledge and understanding then can be tested through the secondary study's datasets. This may take place in the form of comparative secondary research. A study by Tarrant and Hughes (2019) offered an interesting example of working with existing qualitative datasets to compare and illuminate how young fathers were unfavourably positioned by existing policy assumptions and service practices across generations. Using QSA, Tarrant and Hughes (2019) undertook analyses of two datasets that were focused initially on low-income households, family lives, and relationships between different family generations. From the QSA study, the following were found to be common factors across generations of young fathers: the volatility of relationships between young fathers and their partners; insecurity of work and joblessness; nepotism in securing work; difficulties with maternal grandmothers as gatekeepers to their children; and social resources (e.g., employment).

The capacity of QSA to lead to new research studies or data collection is another compelling reason for conducting QSA. In so doing, QSA is combined with a supplementary or independent line of inquiry. Examination of available data using research questions that are different from the original study or with different analytical approaches might reveal gaps and inconsistencies in information collected previously. In this case, conducting a QSA study might involve developing new research questions, refining methodology, and initiating new data collection in order to draw more accurate conclusions (Heaton, 2008; Neale, 2019). The existing data would provide the foundation for a newly designed inquiry that is based on the premise of previously conducted research.

In another study, Watters, Cumming, and Caragata (2018) demonstrated the appealing potential of QSA to combine qualitative data from a prior longitudinal study with new data. In the original study, factors influencing the lives of lone mothers who were supported through social assistance programmes in Canada were explored. Aiming to investigate the role of adversity and lone mothers' resilience that was not studied purposefully in the original research, the collection of additional interview data was initiated. This allowed secondary analysts to enrich the existing dataset with the in-depth accounts and answers to research questions that emerged from the review of the existing data.

In addition to generating new research inquiries and findings, QSA proponents acknowledge the conceptual advancement of QSA as a discrete method (Irwin and Winterton, 2012; Medjedovic, 2011). Secondary analysts striving to broaden and refine QSA as a method can conduct a secondary study to provide more insight into the procedures of original data assessment and analysis based on QSA objectives, data archiving, and QSA ethics (Akcam, Guney and Cresswell, 2019; Bishop and Kuula-Luumi, 2017; Long-Sutehall, Sque and Addington-Hall, 2011). In some instances, secondary research can be used to inform and clarify practices and processes associated with the use of QSA in various social and educational research contexts. Other methodological insights can provide opportunities for novice scholars to engage with the complexities of existing qualitative data and ensure high-quality and trustworthy secondary research.

The growing number of studies in which qualitative secondary analysis is used also reflects changing methodological attitudes towards data archiving. There is a distinctive position on the longevity of data: the longevity of existing data extends beyond the dissemination of research findings and involves data preservation (Beck, 2019; Corti *et al.*, 2014). The activities that extend data longevity, beyond publishing and sharing, include long-term data management and secondary analysis. The increasing availability of studies that use QSA continues to benefit the development of necessary steps in data preservation and gaps in data archiving. A significant example of the importance of re-constructing the context of the original study and archiving procedures was illustrated by Andersson and Ove Sørvik (2013). In two case studies using archived video data from science and language arts classrooms, Andersson and Ove Sørvik (2013) discussed the issues of disclosing context as well as the practices of secondary analysis of existing video data. When conducting QSA, Andersson and Ove Sørvik (2013) highlighted the importance of shared and common procedures for data archiving. These procedures, such as methodological approach, transparency in archiving, and consent for QSA, can help secondary analysts address contextual issues and enable them to access existing qualitative resources to conduct more expansive and comprehensive research.

Together with qualitative data archiving, using QSA leads to valuable reflections on the ethics of the method. Numerous examples of QSA studies can help researchers to make informed decisions about the extent of anonymity, representation, and confidentiality of data. Although these concerns are specific to research contexts and call for consideration within the conceptualisation of a particular QSA study, the changing digital landscape and possibilities of data linkage makes participants in research increasingly recognisable (Edwards, Hughes and Williams, 2015; Hughes and Tarrant, 2019). In a study on childhood poverty, Morrow, Boddy, and Lamb (2014) discussed the complexity of anonymity, as it goes beyond a simple change of names. The authors cautioned that anonymity, in some cases, can remove the significance and diminish the intellectual meaning ascribed to a person's identity. For instance, when asking parents about the history of their child's name, they often referred to a Biblical name that carried a particular spiritual and emotional meaning. As a result, obscuring identities by removing the name can devalue the data essentially and compromise the integrity of the secondary analysis study.

In summary, generation of new knowledge, design of new inquiry, and generation of methodological insights were discussed in this section as being the main QSA objectives, the achievement of which can result in expanded understanding of the topic of interest, its conceptualisation, validation of previously generated research findings, and/or their comparison. Using QSA can also lead to new research questions, data, and/or their expansion. Last, using QSA can contribute to the expansion and refinement of procedures informing data assessment, analysis, archiving, and QSA ethics. However, without systematic considerations for selecting existing qualitative data, trustworthy, complex, and meaningful, QSA becomes challenging.

Considerations for selecting existing data for qualitative secondary analysis

Qualitative data can be used for secondary analyses for different purposes, in different ways, by the original or by different analysts. However, the actual practice of QSA appears to be more challenging compared with the secondary analysis of quantitative data, although revisiting any dataset collected by a third party can be complex and challenging. Several considerations should be taken into account when selecting existing qualitative data for QSA (Sherif, 2018b). The first major consideration is the fit and relevance of existing qualitative data to secondary research, in which secondary analysts conceptualise the direction, objectives, and purpose of QSA study. The second major consideration, quality of the dataset, builds on the challenges associated with the breadth, depth, completeness, and accuracy of existing data. The third consideration, dataset trustworthiness, is about the epistemological and methodological nature of the original study, its context, and background. The fourth consideration, dataset timeliness, involves recognition of the original study's continuum, from its conceptualisation to data archiving, and the recency of existing data within the context of the topic of interest in the present day. A summary of considerations and guiding questions for selecting qualitative data for QSA is presented in Table 4.2.

Within any new QSA study, the focus, sample, and methods of data collection employed in the original research will determine the boundaries of the existing dataset as it relates to initial research questions and study design. As these important dimensions might differ from a QSA study, careful consideration of the relevance and fit of the prospective existing data to the objectives and purpose of the secondary study is warranted. "There [should be] a logical link between the original data set and the question/s asked in the analysis, as the qualitative secondary

Table 4.2 Considerations for selecting qualitative data for secondary analysis

Fit and relevance of dataset to secondary analysis study	• To what degree do existing data, original background, and context relate to the topic of interest? • How relevant are the QSA research questions to the objectives of the original study? • How extensively did participants in the original study report on the topic of interest? • How extensively did participants in the original study experience the topic of interest?
General quality of dataset	• How rich and descriptive are the existing data on the topic of interest? • What data collection methods were used to collect the existing data? Was data triangulation accomplished? • Do secondary analysts, if not the owners of the existing data, have full access to dataset(s) and accompanying documentation? • How sufficient is the dataset documentation (tapes, transcripts, protocols, notes) to fully answer the secondary research question(s) and saturate data? • How sufficient is the dataset documentation (tapes, transcripts, protocols, notes) to re-construct the context of data collection and research settings? • Does the dataset documentation include the type of sample, its size, demographics, and, if possible, geographic descriptors, recruitment, and consent procedures? • If applicable, how accurately were data transcribed/translated? • Does the dataset include transcription protocols with instructions for transcribers and decisions addressing inaudible text segments, overlapping speech, unfamiliar terminology, and language-specific nuances? • Does the dataset include instances of information re-stating, summarising, and/or paraphrasing of participants' insights to ensure the accuracy of the collected data? • Is the access to participant contact details available only to the researcher of the existing data? • When QSA of the existing data is conducted, does the consent form allow for reconnecting with study participants to clarify characteristics of original research and/or to complete missing information?
Trustworthiness of dataset	• How detailed is the description of the original research purpose, questions, and design in the existing dataset? • How detailed is the description of the original study's timeframe, its settings, and data collection settings in the existing dataset? • Does the existing dataset include credentials, institutional affiliations, and contact details of the team members who conducted the original research?
Timeliness of dataset	• Do secondary analysts, if not the owners of the existing data, have full access to the timeline of the research initiation, data collection, and analysis? • Are data points and protocols of data collection time-stamped? • How current and/or relevant are existing data to the present-day topic of interest?

analysis questions [arise] from the original data set" (Du Plessis and Human, 2009, p. 76). More importantly, in order to use existing data appropriately, the available dataset must be adequate to address a proposed research question. When evaluating the fit and relevance of existing data, secondary analysts can consider the following questions: How appropriate is the match between the QSA research questions and those of the existing dataset? How relevant/connected are the QSA research questions to the purpose of the original study? To what extent did the participants of the original study experience the topic of interest?

As secondary analysts assess the fit of the existing dataset, they must ensure that the data are relevant to the objectives and purpose of the QSA study. This requires a thorough understanding of the existing data, the aims and rationale of the original study, the context of the study conducted previously, the data collection methods, and analysis procedures. The existing data must include extensive meta-data (e.g., the original research background, sample type and size, study design, and methods of data collection), which provide critical points of comparison and alignment with the context of the QSA project. The assessment of the dataset fit and relevance defines the answer of whether and how the existing data should and could be used for QSA.

Another consideration when selecting existing data for QSA relates to the quality of the available dataset. Many criteria are included in this consideration and are generally ascribed to the breadth, depth, completeness, and accuracy of the existing data. The data must lead to a thorough understanding and matching of the study settings and people situated within the study sample. Such data provide vivid and rich descriptions of a researched topic of interest while having a strong impact on the findings of the QSA study. When evaluating the quality of the available data, secondary analysts should answer the following questions: How rich and in-depth are the descriptions of participants' experiences, insights, and attitudes to the topic of the QSA study? How sufficient is available documentation to saturate the data and answer secondary research questions? What is the quality of interview/focus group transcriptions and documentation included? The dataset best suited for QSA essentially has the capacity to present every detail on the topic of interest and illustrate how people perceive, react to, and experience it.

Archived, qualitative data that are available for QSA should be examined based on their completeness and accuracy. A complete and accurate dataset has no or minimal missing or damaged data and includes extensive methodological description, field notes, sample and study design decisions, detailed sample plan, and any additional information relevant to the topic of interest and original study organisation and contexts. In addition, secondary analysts should verify the availability of transcription procedures and their timeline. As part of this process, it might be necessary for the researcher to ensure the dataset's credibility via member-checking (Maxwell, 2013) to confirm the accuracy of the existing data.

The trustworthiness of the available dataset must be considered as well. This consideration refers particularly to the comprehensive information about the background of the original study. When assessing the existing data, the dataset should include extensive background information about: (a) the original researchers, data collection methods, and raw data collected (i.e., interviews, interview audio-tapes, and transcripts); (b) characteristics of researcher/s and study participants; (c) data collection site, time of data collection, and data collection settings; and (d) participant recruitment and selection. The following questions can guide the assessment of available data trustworthiness: To what extent does the dataset allow secondary analysts to re-construct and thoroughly understand the nature of the available data? To what extent does the available dataset describe the adherence of the original study to the principles and procedures of qualitative inquiry? As such, there is a possibility that a follow-up with the original researcher(s) might be required to address any concerns or questions.

A final, yet important consideration, when selecting existing data for QSA is the timeliness of the dataset. Because QSA is conducted after the original study, data might be outdated as a result of changing theoretical and conceptual perspectives and/or historical events. If the original research explored time-sensitive questions (e.g., implementation of a policy or programme that is no longer in place, studies on social media) and/or the aim was to portray historical events at the time of data collection, using QSA might fail to lead to an understanding of a research problem in the present day. Although such an outcome might be an opportunity to gain valuable methodological experiences in using QSA, it might fail to lead to an expanded or new understanding of the topic of interest.

Methodological and practical implications of qualitative secondary analysis

In the discussion presented in this chapter, the significance of QSA as a method has been highlighted. The increasing use of QSA in social and educational sciences entails opportunities for the development and refinement of QSA research procedures, including data archiving, sharing, assessment, and analysis. When used in academia, QSA contributes to applied research learning and research development, reducing the need for financial investment and study participant recruitment (see Table 4.3).

Table 4.3 Implications of QSA

Implication	Examples of implications
Methodological	Transparent, ethical, and robust data archiving and sharing; non-intrusive research; continuation of research beyond original study; clear and systematic procedures of existing data selection, assessment and analysis; methodological limitations
Practical	Hands-on instruction; cost-effective research experiences for students and junior faculty; strategies of QSA research design and writing

The use of QSA is increasing as archiving and sharing of qualitative data become more common, transparent, and robust. Further epistemological and methodological development of QSA is also creating opportunities to conduct research on socially important topics, without intrusion into vulnerable populations and disruption of people's daily life. QSA seems increasingly suitable for exploring educational, social, and cultural issues of minorities and marginalised populations. As such, QSA can make a significant contribution to existing knowledge and new theory development without overburdening study participants.

The increasing use of QSA has the potential to advance ethical and archiving guidelines. Extending the discussion on preparation, sharing, and ethical use of existing qualitative data can help researchers recognise QSA as part of the qualitative research continuum that goes beyond the collection and publication of the data for the original study. Ethical reflection and intentional engagement in QSA can help mitigate the ethical dangers of data archiving, sharing, and re-analysis when planning, conducting, and finalising a QSA study. A continuous, critical reflexivity approach to the use of existing data for purposes different from the original research can also help to further understanding of QSA limitations, protect the privacy of study participants, and ensure ethical sensibility in data preservation and QSA research.

QSA also offers practical opportunities for students and junior faculty members. Re-use of existing data can provide hands-on practice to prepare the next generation of qualitative scholars and practitioners. The large number of QSA studies can be used as examples to illustrate and critique the rigour and transparency of original and secondary inquiry, adherence to systematic and comprehensive assessment of existing data selected for QSA, and the limitations of QSA study. Junior faculty members can benefit from the use of QSA as well. Its time-sufficient and cost-effective nature enables junior faculty members to establish their research agenda on the topic of interest, meaningfully contribute to social and educational science research, and collaborate with researchers of existing data.

Conclusions

The aim of this chapter was to provide readers with an overview of discussions about the epistemological and methodological nature of QSA. As a research method that is centred on the re-analysis of existing qualitative data for purposes not established in the original study design, QSA plays a critical role in the development of qualitative research inquiry. Specifically, the aim of using QSA is to generate new knowledge, research inquiry, and methodological insights. As a result, QSA offers an expanded understanding of the topic of interest that was not evident in conceptualisation, validation, and/or comparison of the findings of the original study. The utilisation of QSA influences the emergence of new research questions, extension of the existing datasets, and refinement of procedures related to the QSA ethics, archiving, analysis, and existing data assessment. QSA begins with the identification of a research topic and its supporting data, and is most effective when the selected data are of high quality, sufficient, relevant, and rich. As suggested in this chapter, when selecting an existing dataset, consideration should be given to: the match of the data to the objectives and purpose of a QSA study; data breadth, depth, accuracy, and completeness; comprehensiveness and accessibility of original methodological procedures and decisions; and timeliness of available data. In other words, selection of existing data for QSA should be guided by comprehensive assessment of whether an available dataset allows for developing sufficient understanding of the original study's epistemological and methodological underpinnings, strengths, boundaries, and limitations. In this way, a systematic approach to dataset selection can help secondary analysts to gain the most out of existing data, while accounting for strategies and procedures for ethical QSA research and future data sharing.

Using QSA involves outlining the broader social research and educational context of existing data availability and applicability within which new implications of QSA are arising. Using QSA makes it possible to make greater use of an existing dataset to produce deeper understanding about under-researched phenomena. It can shed more light on experiences of study populations that are challenging to access, while also minimising the emotional discomfort and disruption of their lives. Engaging in QSA must be grounded in ongoing ethical reflection to preserve privacy, anonymity, and confidentiality of data stakeholders.

The educational community might find the use of QSA beneficial as well. Students majoring in Social and Education Studies can use existing data to learn about the complexities of the research process and gain hands-on experiences in qualitative research design, data analysis, archiving, and sharing. Engaging in the assessment of available data provides opportunities for applied learning, where students can further their critical thinking skills and understanding of procedures of ethical research conduct and the importance of methodological transparency. Re-use of qualitative data for teaching and learning purposes encourages the systematic investigation and evaluation of research design and its settings by demonstrating and explaining the research from start to finish. When integrated into teaching, QSA can provide valuable insights into new analytical and theoretical approaches and, as a result, advance the value of existing data to the social and educational sciences field.

References

Akcam, B.K., Guney, S. and Cresswell, A.M. (2019). Research design and major issues in developing dynamic theories by secondary analysis of qualitative data. *Systems*, 7(40), pp. 1–25.

Anderson, S. (2019). *A leadership perspective on family engagement: Qualitative content analysis of secondary data.* Doctoral Dissertation. University of Kentucky, Lexington.

Andersson, E. and Sørvik, G.O. (2013). Reality lost? Re-use of qualitative data in classroom video studies. *Forum Qualitative Sozialforschung / Forum: Qualitative Social Research*, 14(3), Art. 1.

Andrews, M. (2018). Never the last word: Revisiting data. In: M. Andrews, C. Squire and M. Tamboukou (eds.), *Doing Narrative Research*, Sage, London, pp. 205–222

Beck, C.T. (2019). *Secondary Qualitative Data Analysis in the Health and Social Sciences*, Routledge, London.

Bishop, L. and Kuula-Luumi, A. (2017). Revisiting qualitative data reuse: A decade on. *SAGE Open*, 7(1), pp. 1–15.

Carmichael, P. (2017). Secondary qualitative analysis using online resources. In: N.G. Fielding, R.M. Lee and G. Blank (eds.), *The SAGE Handbook of Online Research Methods*, Sage, London, pp. 509–524.

Corti, L. (2012). Recent development in archiving social research. *International Journal of Social Research Methodology*, 15, pp. 281–290.

Corti, L. and Blackhouse, G. (2005). Acquiring qualitative data for secondary analysis. *Forum Qualitative Sozialforschung / Forum: Qualitative Social Research*, 6(2), Art. 36.

Corti, L. and Bishop, L. (2005). Strategies in teaching secondary analysis of qualitative data. *Forum Qualitative Sozialforschung / Forum: Qualitative Social Research*, 6(1), Art. 47.

Corti, L., Van den Eynden, V., Bishop, L. and Wollard, M. (2014). *Managing and Sharing Research Data: A Guide to Good Practice*, Sage, Los Angeles, CA.

Du Plessis, E. and Human, S.P. (2009). Reflecting on meaningful research: A qualitative secondary analysis. *Curationis*, 32(3), pp. 72–79.

Edwards, R., Hughes, C. and Williams, M. (2015). *Data Linkage: Ethical and Social Concerns*. National Center for Research Methods. Available at: https://www.ncrm.ac.uk/news/show.php?article=5444.

Glaser, B.G. (1963). Retreading research materials: The use of secondary analysis by the independent researcher. *The American Behavioral Scientist*, 6, pp. 11–14.

Gupta, S.S. (2010). *State efforts to collect child outcomes data for the Part B-619 and Part C programs under the Individuals with Disabilities Education Act*. Doctoral Dissertation. University of Maryland, College Park.

Gupta, S. and Sherif, V. (2021). Using qualitative secondary analysis to re-examine early intervention leaders' views through a positive lens in a prior systems change. Journal of Interdisciplinary Studies in Education.

Heaton, J. (2000). *Secondary Analysis of Qualitative Data: A Review of the Literature*, Social Policy Research Unit (SPRU), University of York, York.

Heaton, J. (2004). *Reworking Qualitative Data*, Sage, Thousand Oaks, CA.

Heaton, J. (2008). Secondary analysis of qualitative data: An overview. *Historical Social Research*, 33(3), pp. 33–45.

Hinds, P.S., Vogel, R.J. and Clarke-Steffen, L. (1997). The possibilities and pitfalls of doing a secondary analysis of a qualitative data set. *Qualitative Health Research*, 7(3), pp. 408–424.

Holland, J., Thomson, R. and Henderson, S. (2006). *Qualitative longitudinal research*. A discussion paper. London, London South Bank University.

Hughes, K. and Tarrant, A. (2019). *Qualitative Secondary Analysis*, Sage, London.

Irwin, S. (2013). Qualitative secondary data analysis: Ethics, epistemology, and context. *Progress in Development Studies*, 13(4), pp. 295–306.

Irwin, S. (2019). Qualitative secondary analysis: Working across datasets. In: K. Hughes and A. Tarrant (eds.), *Qualitative Secondary Analysis*, Sage, London, pp. 19–36.

Irwin, S., Bornat, J. and Winterton, M. (2012). Timescapes secondary analysis: Comparison, context and working across data sets. *Qualitative Research*, 12(1), pp. 66–80.

Irwin, S. and Winterton, M. (2011). Debates in qualitative secondary analysis: Critical reflections. *Timescapes Working Paper*, 4, pp. 2–23.

Irwin, S. and Winterton, M. (2012). Qualitative secondary analysis and social explanation. *Sociological Research Online*, 17(2), pp. 1–12.

Irwin, S. and Winterton, M. (2014). Gender and work-family conflict: A secondary analysis of timescapes data. In: J. Holland and R. Edwards (eds.), *Understanding Families Over Time*, Palgrave, London, pp. 142–158.

Janzen, K.C. (2012). *Extending the spirit: A qualitative secondary analysis of nurses. Perspectives on spirituality*. Master's Thesis. Trinity Western University, Canada.

Johnston, M.P. (2014). Secondary data analysis: A method of which the time has come. *Qualitative and Quantitative Methods in Libraries*, 3, pp. 619–626.

Kidd, P., Scharf, T. and Veazie, M. (1996). Linking stress and injury in the farming environment: A secondary analysis of qualitative data. *Health Education & Behavior*, 23(2), pp. 224–237.

Long-Sutehall, T., Sque, M. and Addington-Hall, J. (2011). Secondary analysis of qualitative data: A valuable method for exploring sensitive issues with an elusive population? *Journal of Research in Nursing*, 16(4), pp. 335–344.

Mauthner, N.S., Parry, O. and Backett-Milburn, K. (1998). The data are out there, or are they? Implications for archiving and revisiting qualitative data. *Sociology*, 32(4), pp. 733–745.

Maxwell, J.A. (2013). Qualitative Research Design: An Interactive Approach (Applied Social Research Methods), Sage, Thousands Oaks, CA.

Medjedovic, I. (2011). Secondary analysis of qualitative interview data: Objections and experiences. Results of a German feasibility study. *Qualitative Social Research*, 12(3), pp. 1–17.

Mitchell, F. (2015). Reflections on the process of conducting secondary analysis of qualitative data concerning informed choice for young people with a disability in transition. *Forum Qualitative Sozialforschung / Forum: Qualitative Social Research*, 16(3), Art. 10.

Moore, N. (2007). (Re)using qualitative data? *Sociological Research Online*, 12(3), pp. 1–13.

Morrow, V., Boddy, J. and Lamb, R. (2014). The ethics of secondary data analysis: Learning from the experience of sharing qualitative data from young people and their families in an international study of childhood. NOVELLA Working Paper: Narrative Research in Action. Available at: http://eprints.ncrm.ac.uk/3301/1/NOVELLA_NCRM_ethics_of_secondary_analysis.pdf

Neale, B. (2019). Documents of lives and times: Revisiting qualitative data through time. In: K. Hughes and A. Tarrant (eds.), *Qualitative Secondary Analysis*, Sage, London, pp. 61–78.

Sales, E., Lichtenwalter, S. and Fevola, A. (2006). Secondary analysis in social work research education: Past, present, and future promise. *Journal of Social Work Education*, 42(3), pp. 543–558.

Sherif, V. (2018a). Evaluating pre-existing qualitative research data for secondary analysis. *Forum Qualitative Sozialforschung / Forum: Qualitative Social Research*, 19(2), Art. 7.

Sherif, V. (2018b). Practices of youth leadership development in rural high school context: Findings from a qualitative secondary analysis. *Journal of Ethical Educational Leadership: Student Voice and School Leadership*, 1, pp. 273–291.

Smith, E. (2008). Pitfalls and promises: The use of secondary data analysis in educational research. *British Journal of Educational Studies*, 58(3), pp. 323–339.

Stamm, I. (2018). Organized communities as a hybrid from a data sharing: Experiences from the Global STEP Project. *Forum Qualitative Sozialforschung / Forum: Qualitative Social Research*, 19(1), Art. 16.

Tarrant, A. and Hughes, K. (2019). Qualitative secondary analysis: Building longitudinal samples to understand men's generational identities in low-income contexts. *Sociology*, 53(3), pp. 599–611.

Thomson, R. and McLeod, J. (2015). New frontiers in qualitative longitudinal research: An agenda for research. *International Journal of Social Research Methodology*, 18(3), pp. 243–250.

Thorne, S. (2013). Secondary qualitative data analysis. In: C.T. Beck (ed.), *Routledge International Handbook of Qualitative Nursing Research*, Routledge, New York, pp. 393–404.

Vartanian, T.P. (2011). *Secondary Data Analysis*, Oxford University Press, New York.

Watters, E.C., Cumming, S.J. and Caragata, L. (2018). The lone mother resilience project: A qualitative secondary analysis. *Forum Qualitative Sozialforschung /Forum: Qualitative Social Research*, 19(2), Art. 23.

Yardley, S.J., Watts, K.M., Pearson, J. and Richardson, J.C. (2014). Ethical issues in the reuse of qualitative data: Perspectives from literature, practice, and participants. *Qualitative Health Research*, 24(1), pp. 102–113.

5 Evaluation of systematic literature reviews in built environment research

What are we doing and how can we improve?

Vijayan Chelliah, Nicola Thounaojam, Ganesh Devkar, and Boeing Laishram

Introduction

Secondary research is conducted to gather information about a particular domain from a variety of relevant sources. The information is gathered for numerous purposes, particularly to identify the gaps and deficiencies in an area of research, to find out additional information that must be collected, and to form a basis for developing standards and guidelines for practice and policies. This type of research has been used extensively by researchers in the built environment. Moreover, systematic literature review (SLR), also known as systematic review, has demonstrated much attention as a prime tool for carrying out secondary research. A SLR is "a means to identify, analyse and interpret reported evidence related to a set of specific research questions in a way that is unbiased and (to a degree) repeatable" (Kuhrmann *et al.*, 2017). Bias can impact the validity of the results, lead to over-estimation of findings and also make the review untrustworthy for decision-making (Hall *et al.*, 2017). Unlike traditional literature reviews, explicit and rigorous criteria are adopted for SLR, which is used to evaluate critically and synthesise all the literature in a particular domain (Cronin *et al.*, 2008). In addition, SLR differs from other literature reviews because of its "distinct" and "exacting" principles (Denyer and Tranfield, 2009).

With an increasing trend to establish evidence-based practice, researchers are starting to understand the importance of SLRs in built environment research (Parida and Brown, 2018). Considerable improvements in literature review techniques over the last 25 years have shaped SLRs as a powerful research methodology (Denyer and Tranfield, 2009). However, the use of SLR as a secondary research tool has been criticised for its lack of tool support, thus making it challenging to implement, particularly for novices (Kuhrmann *et al.*, 2017).

Researchers face various challenges, for which the available procedures for carrying out SLR do not give sufficient guidance yet (Table 5.1). Often, the best choices of steps to follow in SLR process are unclear. Researchers experience challenges in building adequate search strings to give an accurate set of results. Additionally, they face an extensive body of literature to be screened, classified,

Table 5.1 Challenges in conducting systematic reviews

- How to build adequate search strings to give an accurate set of results? (Booth, 2008; Hall *et al.*, 2017)
- How to control a large body of literature that needs to be filtered and analysed without bias? (Parida and Brown, 2018)
- Which tools and techniques to adopt to analyse the data sets, and finally frame the review results? (Kitchenham, 2004)

Source: Original.

and eventually structured. Also, to organise and analyse the datasets, researchers often have to identify the right tool or method. Therefore, SLRs require appropriate methodological steps to report existing knowledge in an unbiased and structured manner.

The aim of the chapter, therefore, was to present five primary steps that should be followed when conducting SLR. The purpose of the proposed steps is to support scholars in their literature studies by providing a practical and systematic process. Another aim of this chapter also is to review the present status of secondary research in built environment, particularly to evaluate the application of SLR.

The remainder of this chapter has been organised as follows: a description of the recommended steps to design and prepare an SLR; a presentation of the current practices of SLRs in built environment research; and a checklist for authors to adopt an appropriate framework for SLR.

Developing systematic review in built environment research: the five primary steps

Given the concerns and challenges involved in performing a SLR, the notion of SLRs might benefit built environment research through improved quality and diligence, by developing fit-for-purpose SLR methodology. The focus of this section is on discussing several methodological steps that are required when performing an SLR. Modified from the original guidelines, as proposed by Kitchenham (2004), Denyer and Tranfield (2009), the five primary steps are listed in Table 5.2. Each of these steps should be carried out explicitly to obtain compelling and unbiased results.

Table 5.2 The five primary steps for systematic reviews in built environment

1 Formulate research question
2 Locate the literature
3 Select and evaluate the literature
4 Analyse and synthesise the studies
5 Report the review results

Source: Adapted from Kitchenham (2004); Denyer and Tranfield (2009).

The process followed in conducting the review must be documented clearly and explained. Also, justifications for all the steps taken by the researcher must be provided to ensure transparency, which is another key characteristic of SLRs. Researchers often fail to delineate the methods and processes of data collection and analysis in detail, making it difficult to evaluate the quality of the reviews (Denyer and Tranfield, 2009). Including a methodological section, in which the steps performed in conducting the review are described, would enable the readers to determine precisely the scope and quality of the review.

Further, it is important to note that the proposed steps are not meant to be entirely comprehensive. Each of the five primary steps itself could constitute a book chapter. Instead, they have been presented for accessibility to most scholars involved in built environment research and to provide general SLR advice.

Formulate the research question

The first step in an SLR is to formulate the research question or define the subject of the literature review. An SLR should be grounded on a well-defined and answerable research question. However, formulating a research question can be a difficult job, especially for novice reviewers. A common mistake in the beginning is to select a research question that is too broad, making a review unfeasible or too superficial. Therefore, it is always advisable to define the research question narrowly so that the literature generated is manageable (Staples and Niazi, 2007). However, Parida and Brown (2018) argued that a broad approach to the research question should be the first step to view what type of studies and topics are being reviewed currently and finally formulate a concise question.

Lorie and Joan (2014) explored different structures for formulating questions, such as Problem-Intervention-Comparison-Outcome (PICO), Problem-Exposure-Comparison-Outcome-Duration-Results, Person-Environment-Stakeholders-Intervention-Comparison-Outcome, Client Oriented-Practical-Evidence Search, Expectation-Client group-Location-Impact-Professionals-Service, and Population-Interventions-Professionals-Outcome-Health care setting, identified in the health and social science literature. While PICO and the other variations were developed mainly for structuring research questions in the field of health sciences, they are used also in other disciplines with small modification (Davies, 2011). Some of the other question formulation structures developed specifically for qualitative and mixed methods research are Sample-Phenomenon of Interest-Design-Evaluation-Research type and Setting-Perspective-Intervention-Comparison-Evaluation. Similarly, PEO structure is used commonly for qualitative research questions and was constructed based on three concepts: Population, Exposure, and Outcome(s) (PEOs). Researchers may use such tools to formulate concise research questions.

The goals of the search are dictated by the purpose of the review, such as providing an overview of all relevant contributions that address a particular topic. Independent of the respective goals, some of the general research questions particularly worth considering in a literature study are summarised in Table 5.3. Some of these research questions are aimed at giving a demographic summary

Table 5.3 Exemplary standard research questions

- What is the trend of publications (number of articles published over the years) in the particular research domain?
- What are the contributions made under this domain of research?
- What are the key areas of interests under the area of research?
- What are the research gaps?
- What new approaches to topic, methods, materials, and so on are available?

Source: Adapted from Kuhrmann et al. (2017).

of the review and, therefore, will help in preparing to locate and select the literature by checking the fitness of the research questions and modifying them if necessary.

Locate the literature

Having defined the research question, the next step is to locate the literature. To begin this step, researchers should first reflect on appropriate search strings, which are entirely dependent on the research question or scope defined in the previous step. The accuracy of the body of literature identified will depend greatly on the precision of the search strings. Therefore, search strings need careful consideration, and they should always be tested before the actual search. This procedure will ensure also that the review results have taken into account all the available evidence and are based on best quality contributions.

Kitchenham (2004) recommended two strategies to develop the right search strings in advance: snowballing and trial-and-error search. On the one hand, using the snowballing method, the search strings can be tested in advance by studying publications that are already known in the study domain and by extending iteratively the known publication set by checking the references provided. On the other hand, a trial-and-error search is more suitable to finding and testing search queries. In this approach, initial keywords or phrases are required and they are searched using meta-search engines like Scopus or Google Scholar. The purpose here is to examine whether the search query returns a meaningful publication set.

Another issue in locating the literature is with the terminology to use to determine the initial search query. There are technical terms that have hundreds of synonyms and alternative terms. If all of these keywords are not identified, some literature might be missed out of the review, which will affect the quality of the search process. Therefore, it is a good idea to consider synonyms that might generate further information. Booth (2008) recommended adopting a "Building Blocks" strategy in conducting a search query. According to this strategy, the topic of study is broken into facets/blocks. Then, variants and synonyms for each facet are added together using Boolean operators to form a final search query. Some of the most common Boolean operators and their purposes are summarised in Table 5.4. To improve the efficiency and quality of the review, it is wise to spend significant time on building the search query (Denyer and Tranfield, 2009). Further, it is important to mention that different databases follow different search

Table 5.4 Simple operators and Boolean logics for search strings

Operator/logic	Purpose/example
'*'	Refers to the keywords produced by the combination of the root "Sustainab" and any suffix. For example sustainab* searches for literature which contains the term sustainable; sustainability
' " " '	To give the exact phrase. For example "sustainable development"
OR	Searches documents that include any of the keywords identified
AND	Searches documents that include all the keywords identified
NOT	Searches documents excluding the keywords identified

conventions, and hence, reviewers should be cautious about how the operators and Boolean logic are used.

The next step in locating the studies is to identify the databases. Depending on the area of research, several standard sources of data collections are available. Some of the most widely used libraries in the field of built environment are discussed in the next section.

Digital libraries have specific ways of performing the search query construction. So, reviewers should check the appropriateness of the search result by examining whether the set of results contains the expected publications. Also, these libraries have continuous indexing, i.e. the indexes will grow and change over time (Kuhrmann *et al.*, 2017); hence, it is often wise to save the searches in the search history of the database.

Select and evaluate the literature

Reviewing all the literature obtained from the search process is nearly impossible if the sets of results contain a large number of publications. Furthermore, there might be publications that are out of scope or duplicates. Therefore, researchers need to clean the dataset and select the relevant studies. In order to make this procedure scrupulous and reproducible, SLRs require a set of inclusion and exclusion criteria. Some of the inclusion and exclusion criteria that might be useful in selecting an appropriate set of results are listed in Table 5.5. Some of these criteria are found embedded directly into the database and can be used to filter the set of results before exporting the file (e.g. in Scopus researchers can refine the dataset by selecting "limit to" or "exclude" options).

Table 5.5 Exemplary inclusion (I) and exclusion (E) criteria for systematic review

- (I) Title/keywords/abstract of the article make it explicit that the paper is related to (topic/research question)
- (I) Publications available for download
- (E) Publications occurring multiple times (duplicates)
- (E) Publication is not peer-reviewed
- (E) Publication is not in English

Source: Adapted from Kuhrmann *et al.* (2017).

	C	D	E	F	G	H	I	J	K	L	M	N
	Authors	Title	Year	Abstract	Keywords	Source title	Cited by	methodology				Area/Sector
1								Conceptual	Survey	Case study	Literature	
2												
3	Afreen S., Kumar S	Between a rock and a hard	2016	Purpose: Developme	Corporate	Sustainability Ac	3			☑		Port
4	Agostino Cappelli;	Italian Megaprojects: High-!	2010	This paper analyzes the feasibility of a	SCIENZE REGION		0			☑		Transport
8	Basu R.	Managing a major project a	2011	This paper illustrates	Enterprise	International Jou	1	☑	☑	☑		ICT
16	François Molle; Ph	Megaprojects and Social an	2008	Large-scale development of irrigation	Ambio		50			☑		Water
17	Gilmour D., Blackv	Sustainable development ir	2011	The paper presents o	Local	Proceedings of tl	21	☑		☑		Urban
18	Glasson, John	Better monitoring for bette	2005	Drawing on a compre	capacity	Impact Assessme	11			☑		Power

Figure 5.1   Example of minimal data structure required for review (Excel screenshot).

Three of the most widely applied inclusion and exclusion criteria are publication types, multiple occurrence, and language. Researchers should define what types of publications (journals, books, conferences, etc.) are to be included in or excluded from the set of results. Moreover, duplicates are also eliminated. However, it is sometimes difficult to decide which duplicates should be removed. Generally, peer-reviewed journals are regarded as being more up-to-date and rigorous than conference papers (Paternoster *et al.*, 2014).

The dataset obtained through search query needs to be stored in a way in which it can be used for further evaluation. Different databases give different export formats. Therefore, data can be exported in CSV text files or BibTex format, since these formats are more readily converted into spreadsheets. A well-defined data structure should be in place. Figure 5.1 shows a summary of the minimal data structure required to arrange the dataset. This structure can be extended further, depending on the requirements of the study.

Some of the literature obtained in the set of results might consist also of contributions that are out of scope (this scenario can be avoided by defining the search strings appropriately). Therefore, researchers might have to evaluate each of the identified items of literature by title and abstract filtering, if required. The publication selection often includes individual interpretation of a publication, which raises the risk of subjectivity. Therefore, "majority voting" procedure can be selected, which will avoid subjectivity in the publication selection (see detailed discussion in Kuhrmann *et al.* (2017)). Furthermore, in this process, researchers should always remember to align the criteria for inclusion and exclusion with the research questions. Detailed steps should be recorded, specifying precisely the reason for the inclusion or exclusion of publications. The purpose of being precise about these criteria is to make such criteria transparent and reproducible by readers.

Analyse and synthesise the studies

At this point, all the relevant literature will have been collected. The next step is to commence the data analysis and synthesis. The focus of analysis is to get the sense of what the article is about and classify individual studies into some common elements/area of interests. However, the purpose of synthesis is to connect the elements identified from the individual studies and remould them to give overall findings/conclusions/themes/research gaps from the review dataset.

Table 5.6 Exemplary standard question for analysis and synthesis of literature

---

- What type of methodology is adopted (conceptual, survey, case study, literature review)?
- In which contexts (sector, country) was the study conducted?
- What are the broad objectives of the studies?
- What kind of analysis tools is used?
- What are the key findings from the studies?

The first step of analysis is to extract information from every publication included in the review. A series of interrelated questions are often asked to obtain the appropriate information from the review publications. Some of the exemplary questions that can be asked are summarised in Table 5.6.

In addition to answering these questions, it is useful to include the reviewers' key observations and remarks after the item of literature was reviewed. As mentioned above, a "majority voting" procedure should be selected for analysis and synthesis to extract data from studies. Doing this, the interpretations and the findings can be compared, which helps to reduce bias, reconcile any differences, and obtain more robust results. Some of the robust tools that can be used to analyse literature are VOSviewer, Bibexcel, CiteSpace, and methods such as scientometric, bibliometric, infometric, and semantic analysis.

Reporting the review results

Once the analysis and synthesis of the literature have been completed, the review needs to be structured and reported. A good SLR does not end just with a review in a structured and unbiased manner. It should be able also to report systematically the methodology adopted and the findings derived through the review process. These ensure transparency with the readers. Besides, if other reviewers wish to extend or alter the review criteria, they can easily do so.

The structure of the report depends on the purpose of the review or field of study. Nonetheless, some key elements are essential and they should be logical. Fundamentally, the structure includes an introduction, methodology, results, and discussion sections. The introduction section contains the background and research questions. The methodology section contains the precise procedure of how the review was conducted (the key search terms, the databases/libraries used, the inclusion and exclusion criteria, and the analysis and synthesis criteria). Finally, the findings and discussion section includes a summary of all the publications in terms of the information extracted from the articles, such as the trends, sectors, context, and methodologies adopted.

There are several ways in which the findings can be presented, some of which are summarised in Table 5.7. First, researchers can classify the literature into themes or categories. Here, researchers can discuss the distinct themes from the literature through inductive or deductive content analysis. Second, researchers can present the literature in chronological order.

Table 5.7 Ways of reporting the findings from review

Approach	Procedure
Classifying the literature into themes or categories	Thematic or content analysis (inductive or deductive)
Presenting the literature in concepts	Concept mapping
Presenting the literature chronologically	Studies divided into time periods
Exploring the theoretical and methodological literature	Discussion on theoretical literature followed by methodological literature

Source: Adapted from Cronin *et al.* (2008).

Third, the findings also can be presented by exploring the theoretical literature first followed by exploring the methodological literature. This is most likely to be useful when the set of literature is mainly theoretical and can help in identifying the need for qualitative studies. Transparency refers to presenting the findings of the review in such a way that there are clear links between the evidence found and the reviewers' conclusions and recommendations.

What are we doing?

The current practices

Having recommended the five primary steps involved in SLR, a systematic review was undertaken also for this study, using the recommended five steps to provide an overview of the current practices of SLR within built environment research studies. Therefore, the research questions addressed were:

Step 1: Formulate research question (how the research questions in SLRs of built environment research are formulated; how the literature is searched and gathered; how the literature to review is selected; how the items of literature are analysed and synthesised; and how the results of the SLRs are reported). To perform this overview, every article that mentioned "review" (in the title/ abstract/keywords) in 11 journals was retrieved.

Step 2: Locate the literature. The eleven journals were: *Building and Environment; Engineering, Construction, and Architectural Management; Automation in Construction; International Journal of Civil Engineering and Technology; Journal of Management of Engineering; Construction Innovations; Construction Management and Economics; Journal of Building Engineering; Journal of Construction Engineering and Management; Built Environment Project and Asset Management;* and *International Journal of Construction Management.* From this search, 242 articles were discovered. The articles were then selected based on the review methodology, i.e. only those articles that performed SLR were selected.

Step 3: Select and evaluate the literature. Out of 242 review articles, 54 articles reported an SLR. Despite an increase in importance of SLRs in built environment, it was found from the review that only 22% of literature reviews performed systematic literature reviews. Parida and Brown (2018) argued also that the SLR technique in the built environment was infrequent. This might be because of the challenges reviewers often face in performing SLRs. Often, the appropriate steps to be taken in SLR process are unclear and there is a lack of sufficient guidance. The selected articles for review were exported into Excel with primary details of each article included.

Step 4: Analyse and synthesise the studies. Content analysis of the selected review papers was performed to address the research questions. The aim of this step was to assess whether the quality of these 54 SLR studies was appropriate (in terms of the recommended steps). The aim was to explore and see how these steps were implemented in the reviews conducted by researchers in the field of built environment. The findings of the review are reported in the following sub-sections.

Step 5: Report the review results.

Formulating research question

Few researchers (20 articles) have attempted to mention the references of the approach to carrying out the systematic review. Some of the approaches adopted were the steps proposed by Kitchenham (2004), Pawson *et al.* (2005), Higgins and Green (2008), Denyer and Tranfield (2009), Moher *et al.* (2010), Okoli (2015), Thomé *et al.* (2016). These approaches are described briefly in Table 5.8. Also, researchers dictated the aim and goals of the reviews in generic terms. However, authors failed to follow question formulation structures such as PICO or PEO (discussed earlier) to structure a research question that would facilitate a more focused search. Only one article out of 54 was found to have adopted PICO to formulate the research question.

Locate the literature

The search string is one of the most crucial elements because its precision will affect the set of results. However, most of the reviewers failed to reflect on appropriate search strings. Even though authors disclosed the keywords selected, they failed to provide the reasons for the selection of particular keywords. Researchers also omitted the use of alternative vocabulary in forming the search query. This could lead to the exclusion of some relevant studies that are part of the research topic or question. In terms of using Boolean logic to form the search query, the review found that authors used mainly Boolean logic. However, it appears that the authors were not much aware of the importance of using truncating operators. As for the databases selected for searches of publications, researchers used mostly Scopus, followed by Web of Science (WoS), as shown in Figure 5.2. Authors also used libraries such as ASCE, Emerald, T&F, and Elsevier. Moreover, authors chose publications based on some selected journals. For instance, few researchers limited their articles from the journal list ranked by Chau (1997). Out of 54 articles,

Table 5.8 Some approaches to SLR process followed in built environment research

Approaches	Steps
Kitchenham (2004)	1 Identify research 2 Study selection 3 Study quality assessment 4 Extract data 5 Synthesise data
Pawson *et al.* (2005)	1 Clarify scope 2 Search for evidence 3 Appraise primary studies and extract data 4 Synthesise evidence and draw conclusions 5 Disseminate, implement, and evaluate
Higgins and Green (2008)	1 Determine scope and research questions 2 Define inclusion criteria for including and grouping of studies for synthesis 3 Search and select studies 4 Assess risk of bias 5 Synthesise, meta-analyse, and interpret results 6 Present findings
Denyer and Tranfield (2009)	1 Formulate question 2 Locate studies 3 Select and evaluate study 4 Analyse and synthesise 5 Report and use results
Moher *et al.* (2010)	1 Describe rationale and objectives 2 Specify eligibility criteria 3 Describe intended information sources 4 Present search strategy 5 Show study records (data management, selection process, data collection process) 6 List data items 7 Assess risk of bias 8 Synthesis of data
Okoli (2015)	1 Identify the purpose 2 Draft protocol and train team 3 Screening for inclusion 4 Search for literature 5 Extract data 6 Appraise quality 7 Synthesise studies 8 Write the review
Thomé *et al.* (2016)	1 Plan and formulate the problem 2 Search the literature 3 Gather data 4 Evaluate quality 5 Analyse and synthesise data 6 Interpret analysis 7 Present results 8 Update the review

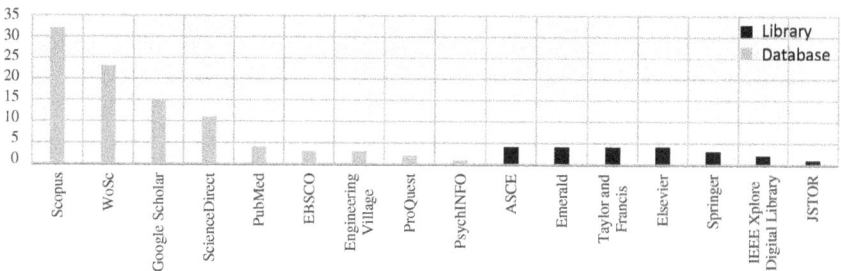

Figure 5.2 Databases and libraries used in built environment reviews.

16 included attempts to provide the rationale for choosing a particular database or library. In addition, 17 articles were found for which the publications had been selected from only one database. However, Bramer *et al.* (2017) argued that as a "minimum requirement", searching from a combination of databases such as Scopus, WoS, and Google Scholar can ensure acceptable and well-ordered coverage of publications.

Select and evaluate the literature

From the review it was found that most of the researchers specified the inclusion and exclusion criteria, excluding predominantly publications that were not peer-reviewed, publications that were duplicated, and publications that were not in English. Nonetheless, most of the reviews did not have specified reasons for including or excluding literature. Providing the rationale for including or excluding a publication would enable readers to understand how the criteria were aligned to the research question and, hence, ensure transparency. In addition, it was found that the concept of "majority voting", which should result in more objectivity in the selection of publications, had not been used by any reviewers.

Analyse and synthesise the studies

As part of analysing and synthesising the studies, researchers were found to have used tools such as Excel, VOSviewer, CiteSpace, Gephi, and NVivo. Also, authors used the support of methods such as meta-analysis, scientometric, constant comparative analysis, and content analysis. Predominantly, researchers presented their analysis of the studies according to a number of publications per year, type of methodologies adopted, and a number of studies in major research categories.

Reporting the review results

How the review results are reported depends entirely on the motive of the study or field of study. Generally, researchers reported their reviews in terms of emerging trends of research, contributions, classification of methods, and tools and techniques used in the studies. Moreover, authors identified the major research gaps and areas for development in a particular domain. To review articles that had specific research objectives such as to find parameters, factors or criteria, authors used methods such as content analysis and concept mapping to develop frameworks. To conclude the chapter a checklist of steps to be taken in producing a sound, unbiased and structured SLR has been provided in the next section.

How can we improve?

A *checklist*

Overall, it was found that several challenges existed in performing SLR, and reviewers in the field of built environment also failed to follow some of the recommended steps. Table 5.9 has been included to summarise the recommended best practices and to provide a checklist for researchers. It is recommended that researchers performing SLRs should follow this checklist, step by step, to ensure order and transparency without bias in performing and reporting SLRs.

Conclusions

SLRs have become an important means to assess and structure the context of the existing body of knowledge. In built environment studies, researchers are starting to understand the importance of SLRs. However, when conducting such a review, there are various challenges, mostly concerning lack of tool support or insufficient guidance.

In this chapter, five primary steps that should be followed in the process of undertaking SLRs have been provided. The five steps are: (a) formulate research question, (b) locate the literature, (c) select and evaluate the literature, (d) analyse and synthesise the studies, and (e) report the review results. Also, all of the five primary steps of SLRs were reviewed in this chapter, in the process of reviewing 11 journals in the area of built environment.

From the results it was found that a SLR had been performed for only 54 out of 242 studies. Under research question formulation, researchers failed to follow question formulation structures that would facilitate a more focused search. Furthermore, most of the reviewers failed to use alternative vocabulary in forming search queries and also omitted the rationale for selection of keywords. As part of database selection, the results indicated that researchers used mostly Scopus and WoS. A combination of databases is recommended to ensure efficient and acceptable coverage. However, a few articles used only one database source. Finally, in this chapter, a step-by-step guide or checklist has been provided to enable researchers, especially novices, to design better systematic reviews.

Table 5.9 Proposed checklist of recommendations for systematic review

SLR main steps	SLR sub-steps	Recommendations	Checklist
1 Formulate research question		Did you follow any question formats to structure a research question that will facilitate a focused search?	☐
2 Locate the literature	2.1 Search string	What alternative synonym is used in defining your topic? Are there variants in terms of American and British spelling? Is there a word-stem for truncation, for example child* to find child, children, or childish? Are acronyms and abbreviations considered? In what specific examples, cases, and terms are you interested? Are there other groups/classes you would like to exclude?	☐ ☐ ☐ ☐ ☐ ☐ ☐ ☐ ☐ ☐ ☐ ☐
	2.2 Search strategy	Did you use the Boolean operation with the search string?	
	2.3 Select database and literature coverage	Did you perform a literature search in a minimum of three of these databases: Scopus, EBSCO, WoS, Google Scholar, and Science Direct?	
3 Select and evaluate the literature		Did you follow a protocol to select and evaluate the literature in terms of document type, period, language, and so on? Did you keep record of all steps followed? Did you specify the inclusion and exclusion criteria clearly? Did you specify the reason for inclusion and exclusion of literature?	☐ ☐ ☐ ☐ ☐ ☐ ☐ ☐
4 Analyse and synthesise the studies		Did you use an appropriate method/technique and robust tools to analyse the literature? Such as: - Methods: scientometric, bibliometric, informetric, semantic analysis, and others - Tools: VOSviewer, Bibexcel, Citespace, and so on	
5 Report the review results		Did you disseminate the findings in themes/categories using a guideline, framework, or a theory?	☐

References

Booth, A. (2008). Unpacking your literature search toolbox: On search styles and tactics. *Health Information & Libraries Journal*, 25(4), pp. 313–317.

Bramer, W.M., Rethlefsen, M.L., Kleijnen, J. and Franco, O.H. (2017). Optimal database combinations for literature searches in systematic reviews: A prospective exploratory study. *Systematic Reviews*, 6(1), pp. 245–257.

Chau, W.K. (1997). The ranking of construction management journals. *Construction Management and Economics*, 15(4), pp. 387–398.

Cronin, P., Ryan, F. and Coughlan, M. (2008). Undertaking a literature review: A step-by-step approach. *British Journal of Nursing*, 17(1), pp. 38–43.

Davies, K.S. (2011). Formulating the evidence based practice question: A review of the frameworks. *Evidence Based Library and Information Practice*, 6(2), pp. 75–80.

Denyer, D. and Tranfield, D. (2009). Producing a systematic review. In: B.D. Bryman and A. Bryman (eds.), *The Sage Handbook of Organizational Research Methods*, Sage, London, pp. 671–689.

Hall, S., Oldfield, P., Mullins, B.J., Pollard, B. and Criado-Perez, C. (2017). Evidence based practice for the built environment: Can systematic reviews close the research - practice gap? *Procedia Engineering*, 180, pp. 912–924.

Higgins, J.P. and Green, S. (2008). *Cochrane Handbook for Systematic Reviews of Interventions*, 1st edition, John Wiley & Sons, Chichester.

Kitchenham, B. (2004). *Procedures for Performing Systematic Reviews*, Keele University, Keele.

Kuhrmann, M., Fernández, D.M. and Daneva, M. (2017). On the pragmatic design of literature studies in software engineering: An experience-based guideline. *Empirical Software Engineering*, 22(6), pp. 2852–2891.

Lorie, K. and Joan, C.B. (2014). Formulating answerable questions: Question negotiation in evidence-based practice. *Journal of the Canadian Health Libraries Association*, 34(2), pp. 55–60.

Moher, D., Liberati, A., Tetzlaff, J. and Altman, D.G. (2010). Preferred reporting items for systematic reviews and meta-analyses: The PRISMA statement. *International Journal of Surgery*, 8(5), pp. 336–341.

Okoli, C. (2015). A guide to conducting a standalone systematic literature review. *Communications of the Association for Information Systems*, 37(43), pp. 879–910.

Parida, S. and Brown, K. (2018). Investigating systematic review for multi-disciplinary research in the built environment. *Built Environment Project and Asset Management*, 8(1), pp. 78–90.

Paternoster, N., Giardino, C., Unterkalmsteiner, M., Gorschek, T. and Abrahamsson, P. (2014). Software development in startup companies: A systematic mapping study. *Information and Software Technology*, 56(10), pp. 1200–1218.

Pawson, R., Greenhalgh, T., Harvey, G. and Walshe, K. (2005). Realist review – A new method of systematic review designed for complex policy interventions'. *Journal of Health Services Research & Policy*, 10(1), pp. 21–34.

Staples, M. and Niazi, M. (2007). Experiences using systematic review guidelines. *Journal of Systems and Software*, 80(9), pp. 1425–1437.

Thomé, A.M.T., Scavarda, L.F. and Scavarda, A.J. (2016). Conducting systematic literature review in operations management. *Production Planning & Control*, 27(5), pp. 408–420.

6 When does published literature constitute data for secondary research and how should the data be analysed?

Saad Sarhan and Emmanuel Manu

Introduction

The accumulation of empirical studies conducted in the built environment has led to a growing body of literature and vast amounts of data that have been collected, collated, and archived by researchers from all over the world. Whilst attention has been focused on the processes and challenges of obtaining valid and reliable primary data in research studies, not enough focus has been given to the use of secondary research methods in the built environment. Secondary research involves the use of existing data obtained by others for a purpose that can be different from that for which they were collected originally. Whilst the focus of some secondary methods, such as qualitative secondary analysis (see Chapter 4), is on re-using the original qualitative data from primary studies, there are several secondary methods that can be used to analyse the outcomes of primary studies (as presented in published academic literature) in an objective and unbiased manner, using either quantitative or qualitative forms of synthesis. The utilisation of published literature as secondary data not only provides a viable alternative option for researchers facing time, resource, and cost constraints, but also can be used to supplement primary data, whilst considering both the benefits and shortcomings of doing so.

One of the common forms of secondary research is systematic literature review. With the growing concern about the quality and variability of research evidence being used to inform teaching, practice, and policy in the built environment, systemic literature reviews have been used increasingly by built environment researchers to identify, evaluate and synthesise the outcomes of all relevant empirical studies on a certain topic. In addition, the advancement in technology and software has enabled researchers to analyse and deal with large amounts of data more efficiently. The purpose of this chapter, therefore, is to differentiate the traditional narrative review from established secondary research methods, in which published literature is used as secondary data, before providing methodical guidance on how to analyse existing academic literature systematically as part of a qualitative secondary research method (in this instance, a systematic literature review). This chapter also includes some discussion on the use of qualitative data analysis software (i.e. NVivo) to produce a qualitative synthesis of the published literature when conducting systematic literature reviews. Whilst there are many published studies that focus on describing how NVivo can be used to analyse qualitative data, an illustrative example is used in this chapter to present a coherent methodical approach for conducting systematic literature reviews using NVivo.

Traditional/narrative literature review

It is a well-established academic convention that the first research step undertaken by most post-graduate and doctoral students is to review literature relevant to their chosen topic of study. This review of the literature is often conducted using a traditional or narrative approach, which is illustrated in Box 6.1 and helps to provide the background needed to understand current knowledge. The literature review enables students to become familiar with the current trends, theories, methods, and key scholars in their field of research. It also enables them to identify gaps or inconsistencies in knowledge that they could explore and address through their own research. A literature review can be described as an objective, thorough summary and critical analysis of the relevant, available literature on the topic of interest being studied (Hart, 1998). The main purpose of the review is to provide the reader with a comprehensive and detailed overview of current literature on a specific topic, and to form the basis for other goals, such as offering recommendations for future research in the area (Cronin *et al.*, 2008). Reviewing literature can inspire research by identifying trends, gaps, and inconsistencies in an existing body of knowledge, thus helping the researcher to develop research questions, hypotheses, conceptual models, or frameworks.

However, reviewing the literature can be a demanding undertaking that requires learning and developing complex skills, ranging from defining topics for exploration, searching and retrieving literature, critical thinking and citation skills, analysing and synthesising data from various sources, to being proficient at paraphrasing and writing, often within a limited time scale (Pautasso, 2013). The traditional or narrative approach to reviewing literature requires that students read and examine as much of the relevant literature as they possibly can, and then construct a carefully argued narrative of their research analysis (Pickering and

---

BOX 6.1 TRADITIONAL LITERATURE REVIEW PROCESS

A literature review process typically comprises the following:

- Identifying a review topic;
- Searching the literature;
- Gathering, reading, and analysing the literature;
- Synthesising the findings;
- Writing the review;
- Citing and listing the references and bibliography.

Byrne, 2014). Often, aspects of these traditional or narrative literature reviews are presented as part of the literature review chapter in dissertations and theses. This traditional narrative approach to literature review is common in the academic literature of a wide range of disciplines, including those of the built environment.

Given that the review of literature is a key step in any research, be it primary or secondary, there are key questions that require clarification regarding the use of traditional narrative reviews and how this relates to secondary research methods in which published literature is analysed. These questions are addressed below:

- Should the literature review conducted using the traditional narrative approach be reported as an application of a secondary research method?

Students and researchers who are not familiar with the use of secondary research methods are often tempted to report their narrative review of the literature, conducted as a starting point of their study, as an application of a secondary research method. This is not entirely accurate because, if it were, then that would mean that all research is based on secondary methods just because literature had to be reviewed to identify gaps or inconsistencies in knowledge, and to define the research questions, scope, and focus of the research. So, conducting a traditional literature review as a starting point of all research does not equate to the application of a secondary research method.

- When using published literature as part of a secondary research method, will it still be necessary to conduct and report a traditional literature review as part of the study?

Yes, because even when applying a secondary research method for any research, the first step of the process is still to review the literature to identify gaps in the research and formulate the research questions that will drive the rest of the study (see Chapter 1). This initial literature review is often based on the traditional narrative approach and is reported as the literature review conducted as part of the secondary research study.

- At what point does the use of published literature become a secondary research method in its own right?

Beyond the traditional review of literature as part of any research, there are secondary research methodologies in which published literature is used as secondary data which will be analysed to arrive at findings, just as in the analysis of primary data that is conducted in primary research. These methods, which are summarised in Table 6.1, are sometimes classified broadly as systematic reviews, and include systematic literature reviews or meta-synthesis, scoping reviews, state-of-the-art reviews, bibliometric review, and meta-analysis.

Systematic literature reviews, scoping reviews, and state-of-the-art reviews are all secondary methods that are mainly based on a qualitative strategy of inquiry. Of these qualitative approaches, systematic literature reviews seem to be the most popular method that has been applied in built environment research. Scoping reviews are a relatively new method that can also be used to synthesise and provide

Table 6.1 A summary of secondary research methods that use published literature as data for analysis

Type of review	Description
Systematic literature review or meta-synthesis	A systematic approach that is used to integrate, evaluate, and interpret the findings of published research studies. This approach is used to identify themes or concepts that exist within or across individual qualitative studies. The researcher may use comprehensive searching according to explicit inclusion/exclusion criteria or use purposive sampling
Scoping review	A relatively new method for providing evidence of literature synthesis. This method differs from systematic literature reviews in terms of its purpose and aim. The aim is to identify and assess the nature and extent of available research literature on a specific topic, usually including ongoing research. Completeness of searching is determined by scope or time constraints
State-of-the-art review	An approach used to summarise the current state of knowledge and provide priorities for future research investigations. This approach is focused on current and recent rather than retrospective issues. The aim is to determine comprehensiveness of current literature rather than existing literature
Bibliometric review	A systematic research approach in which quantitative methods and statistics are used to analyse the bibliographic data of published literature. The approach is used often to measure the impact of research authors, publications, or topics within a given subject area or body of literature
Meta-analysis	A review technique used to combine statistically the results of prior quantitative studies to achieve more precise results and to enhance their objectivity and validity. The aim is comprehensive searching. Completeness may be assessed by using funnel plot examination

Source: Adapted from Grant and Booth (2009).

evidence of findings using published literature, but differs from systematic literature reviews in terms of its purpose and aims (Sucharew and Maurizio Macaluso, 2019). Scoping review as a secondary method is still yet to be applied extensively in built environment research.

Bibliometric reviews or analysis and meta-analysis are systematic reviews that utilise more quantitative approaches. Bibliometric reviews are based on statistical analysis of the bibliographic data in published literature, and their application in built environment research has been on the ascendancy in recent years. Meta-analysis is a systematic review that is used to combine the results of prior quantitative analysis so that these can be analysed further at a higher analytical level. Some of these methods are discussed more extensively in other chapters of this book (see Chapters 9-13 for bibliometric approaches and Chapter 15 for an application of meta-analysis). In the rest of this chapter, the systematic literature review method is discussed in more detail, based on which the approaches for analysing the published literature qualitatively, as secondary data are discussed.

Systematic literature review as a form of secondary research

In contrast to the traditional narrative literature review, in which *ad hoc* literature selection is used to describe observed features and is subject to unconscious bias, systematic literature review is based on a more explicit, defined, and rigorous approach to reviewing what is deemed to be all relevant studies on a specific topic. In other words, a systematic literature review is used to identify, critically evaluate, and synthesise the findings of all relevant, high-quality individual studies that fit pre-specified eligibility criteria to address one or more research questions. A rigorous and transparent methodical approach is used throughout the entire research process in order to reduce bias and enable future replication (Sarhan *et al.*, 2019). By combining the evidence from multiple studies, systematic literature reviews help to create a "greater whole" or develop a "bigger picture" of a specific topic (Pickering and Byrne, 2014). A systematic literature review (see process in Box 6.2), therefore, is regarded as a research project on its own merits, which can be used to address much broader questions than any single empirical study ever can (Siddaway *et al.*, 2019).

The systematic approach to searching, selecting, and synthesising published literature as qualitative secondary data can be performed manually or supported with qualitative data analysis software applications such as NVivo. The academic literature that has been systematically identified is treated as interview transcripts by the researcher. The selected academic literature is then analysed as a form of secondary data, using qualitative approaches to research synthesis. It is worth noting however that being systematic does not mean just being comprehensive in literature searching or simply using explicit inclusion/exclusion criteria to control the size of the literature sample (Badger *et al.*, 2000). What is most important is to follow a coherent, methodical approach during the entire research and to retain a logic of the decisions made in the process. In the following sections, therefore, an overview of the main methods and processes used for data analysis when applying systematic literature reviews is provided.

BOX 6.2 SYSTEMATIC LITERATURE REVIEW PROCESS

Systematic literature reviews typically require the researcher explicitly to:

- formulate the research question(s) based on an initial review of literature;
- set inclusion and/or exclusion criteria;
- search, access, and identify all the relevant literature;
- assess the quality of the literature to be included in the review;
- analyse, synthesise, and disseminate the findings;
- cite and create a list of the literature sample being reviewed.

Approaches to data analysis in systematic literature reviews

There are two approaches to data analysis commonly used by researchers in qualitative descriptive research and systematic reviews, namely: (1) Thematic Analysis (see Braun and Clarke, 2006; Guest *et al.*, 2012); and (2) Qualitative Content Analysis (see Hsieh and Shannon, 2005; Elo and Kyngas, 2008). Other approaches include the use of grounded-theory methodologies for data coding and analysis (see Glaser and Strauss, 1967; Strauss and Corbin, 1998; Charmaz, 2008). Table 6.2 shows a summary of the main characteristics of thematic and qualitative content analysis, which are presented in the form of a comparison.

Thematic analysis can be defined as a method used for systematically identifying, organising, and providing insight into patterns of meaning (themes) across a dataset (e.g. texts, transcripts, and literature). The method provides researchers with a flexible, albeit well structured, way of reviewing and making sense of what has been said or written about a specific topic (Nowell *et al.*, 2017; Neuendorf, 2019), thereby helping researchers to identify patterns and generate themes from the data. A "theme" can be described as a concept that captures something important about the data in relation to the research question (Vaismoradi *et al.*, 2016). A theme represents some level of patterned response or underlying meaning within the dataset. Each theme may also have some sub-themes, which help to provide a comprehensive or contextual view of the patterns in the coded data. Braun and Clarke (2006; see also Braun and Clarke, 2012, pp. 57–71) provide a six-phase recursive process for conducting thematic analysis, which can be summarised as follows:

1   Familiarise yourself with the data (e.g. transcribe data if necessary, reading and reviewing the data, and taking notes about ideas and questions).
2   Generate initial codes – code data in an inductive manner (see procedures for data coding in the following section of this chapter) and collate relevant data to each generated code.

Table 6.2 A comparison between qualitative content analysis and thematic analysis

	Qualitative content analysis	*Thematic analysis*
Reporting style	A simple report of common trends in data	Provision of a rich and complex account of data
Data analysis	Both qualifying and quantifying data	Pure qualitative data analysis
Coding	Uses frequency of codes to find meanings – be wary of the danger of capturing the surface meaning	Higher level of interpretation than description – captures the core of the phenomenon
Limitation	Danger of missing the context and the hidden agenda	Theoretical flexibility can lead to inconsistency and a lack of coherence

Source: Adapted from Vaismoradi and Snelgrove (2019).

3 Search for themes – examine the codes and collate them into potential themes to capture important patterns of meaning.
4 Review themes – check whether the developed themes work in relation to the coded data and generate a thematic map.
5 Define and name themes – perform ongoing analysis to refine the specifics, names and definitions for each theme, and ensure that they provide an interesting story.
6 Produce a report – reflect on the overall analysis and produce a scholarly report of the thematic analysis.

However, "content analysis" is a general term that is used for a number of different approaches to analysing textual data (Vaismoradi *et al.*, 2013). In its most common quantitative form, which historically follows a positivist paradigm (Berelson, 1952), initial codes are developed beforehand by the researcher in a deductive manner (Neuendorf, 2019). This means that the researcher approaches the data with a series of concepts, ideas, or topics in the form of a priori codes. The researcher then analyses the texts to find relevant data that could be collated with the predetermined codes, i.e. a priori coding as opposed to emergent coding (Blair, 2015). A typical example can be seen in the use of content analysis by Mzyece *et al.* (2019) to synthesise data on information production or exchange from the Construction (Design and Management) regulations 2015. When applying content analysis, it is mainly the frequency of codes that is counted to indicate significant patterns in the text, which then influences the development of themes. There is, however, the danger of missing the context and not capturing the core issues of the phenomenon under investigation. For this reason, amongst others, more recent variations have introduced "qualitative content analysis" (Graneheim *et al.*, 2017; Neuendorf, 2019). This approach was applied for example by Brewer and Strahorn (2012), who used qualitative content analysis to analyse the dimensions of trust that are reflected in the Project Management Body of Knowledge (PMBOK).

In qualitative content analysis, it is possible to analyse data in both qualitative and quantitative ways, but in thematic analysis, only a pure, qualitative data analysis process is used. In the tradition of quantitative data analysis, the quantification of data means transforming words and codes into numbers for data analysis. In contrast, quantifying data within the context of qualitative content analysis implies that the frequency of codes is important, but as an intermediary step in data analysis, for developing and determining the importance of a theme. In qualitative content analysis, qualitative techniques are used hand-in-hand with code frequency to describe, interpret, or explore meanings of data in more detail. They also can be used together to verify the findings, a technique known as "triangulation" in mixed methods research (Vaismoradi *et al.*, 2019).

In both thematic analysis and qualitative content analysis the aim is to identify and develop patterns or themes based on rigorous procedures for data coding and analysis. However, the decision about which method to use depends mainly on the nature of the study (e.g. research question and objectives) and the researcher's motivation in the analytical process. The decision includes whether the researcher aims to reach descriptive (manifest content) or inter-pretive (latent content) levels of analysis. In thematic analysis, researchers rely on the underlying meanings of codes throughout the entire dataset as a means for developing themes. A theme would be considered to be important if it is believed to provide important insights to the research question. However, researchers using qualitative content analysis tend to rely primarily on code frequency for finding and developing themes. It can also be suggested that the-matic analysis is conducted mainly in an inductive manner, while qualitative content analysis can be used in deductive or inductive ways (Elo and Kyngas, 2008). In reality, a combination of both approaches is often used; however, one approach tends to predominate (Braun and Clarke, 2012). Therefore, it is important for researchers to understand the reasoning stances that they take, to ensure consistency and coherence of their overall approach to data analysis. Further guidance is provided in the following section on how to analyse data (i.e. literature) qualitatively, with a particular focus on data coding techniques.

Processes for analysing qualitative data

Data analysis, in qualitative research, entails systematically organising and mak-ing sense of qualitative data that are collected by the researcher to gain a better understanding of the phenomenon under investigation (Figure 6.1). The process of analysing qualitative data involves coding large amounts of data, which can

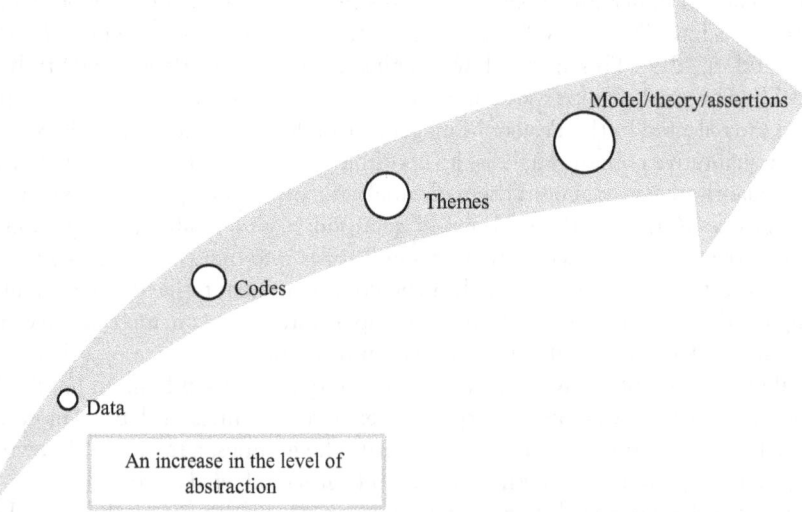

Figure 6.1 Processes and levels of abstraction in qualitative data analysis.
Source: Original.

include academic literature, interview transcripts, observational notes, document reviews, or other non-textual material (e.g. pictures and videos). Codes are tags or labels attached to chunks of data of varying size, which could be words, phrases, sentences, or whole paragraphs.

Coding will depend on whether the themes being developed are more theory-driven, following deductive approaches, or data-driven, using inductive approaches. In the former, themes would be pre-determined and the researcher would code data that fit with the themes, i.e. a "top-down" approach to theorising. While in the latter, themes would be emergent as coding would be conducted in an inductive manner without the researcher being influenced by prior literature or experience, i.e. a "bottom-up" approach to theorising. In other cases, it might be useful for the researcher to use or integrate both approaches (Ali and Birley, 1999). For example, in a systematic literature review, Sarhan *et al.* (2019) adopted a "deductive-inductive" reasoning approach to data coding and analysis (Figure 6.2) to explore the synergies and inconsistencies between "Lean and Sustainable Construction" theories and practices. In the study, Sarhan *et al.* (2019) followed a "lean coding" approach (Creswell, 2007, p. 152) as opposed to purely inductive coding approaches, where researchers might struggle to reduce the numerous lists of generated codes to the five or six main categories or themes that they should arrive at for most publications. In lean coding, the researcher would start the data-coding process by developing a shortlist of five or six main themes with shorthand codes, and then expand and refine the coding structure inductively as they continue to review the literature.

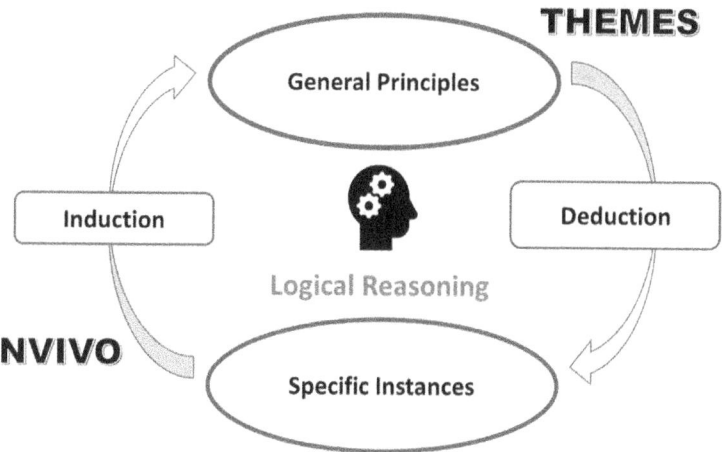

Figure 6.2 A deductive-inductive (integrated) reasoning approach for data coding and analysis.

Source: Original.

Coding requires researchers to make sense out of the data being analysed. This includes asking themselves the following questions:

- What is this data saying?
- What does this represent?
- What is this an example of?
- What kind of events are at issue here?
- What do I see is going on here?
- How has this happened?
- Why has this happened?
- What is trying to be conveyed?

Details of the process that should be followed when analysing qualitative data, be it for primary or secondary qualitative data, are summarised in Box 6.3.

Manual or computer-assisted methods for qualitative data analysis?

Traditionally, data coding and categorising were conducted manually with the use of coloured pens, papers, note cards, and a pair of scissors to mark, cut, and sort the data (Figure 6.3). Other traditional processes for qualitative data analysis included manually counting the frequency of words or collections of words (King, 2008). In general, such manual processes could be very time-consuming and complicated.

Over recent years, researchers have increasingly made use of computer-assisted qualitative data analysis software (CAQDAS). The use of CAQDAS is regarded as being significant in terms of increasing the efficiency and speeding up the process of categorising and retrieving coded data, something that previously has been daunting to perform manually (King, 2008; Salmona and Kaczynski,

---

BOX 6.3 QUALITATIVE DATA ANALYSIS PROCESS

Qualitative data analysis typically involves the following iterative processes:

- Transcribing all data (in systematic literature reviews, this will be the published literature selection that will be analysed as secondary data).
- Coding the data. Codes serve as a way to label, organise, and sort your data.
- Searching for significant patterns in the data (e.g. similarity, difference, frequency).
- Identifying emergent themes or categories.
- Development of higher order themes or conceptual model.
- Review/evaluation of developed themes or model.

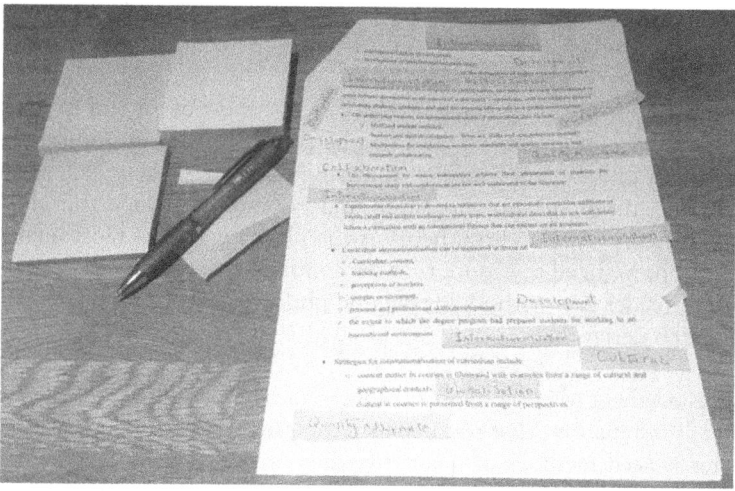

Figure 6.3 Manual methods for qualitative data analysis.
Source: Original.

2016), especially when dealing with large amounts of qualitative data. CAQ-DAS is regarded also as being useful in aiding the researcher to keep records, and in providing an audit of the entire data analysis process, thereby enhancing the transparency and quality of the research (Welsh, 2002; Bringer *et al.*, 2004). Nevertheless, it is important to stress that CAQDAS programmes do not analyse the data for the researcher. It is the researcher who decides what to code, collate, and categorise, and who draws meanings from the data rather than the software programme used.

There are some concerns and downfalls associated with the use of CAQDAS. For instance, the time investment needed in order to learn and become familiar with the CAQDAS package is a significant factor worth considering (Welsh, 2002; Salmona and Kaczynski, 2016). That is because CAQDAS might not be appropriate for use when dealing with small amounts of data (The University of Northampton, 2017). There is also the danger of researchers "getting too close to the data" without allowing the distance needed to reflect and consider the bigger picture (King, 2008), a phenomenon that is described as the "coding trap" (Gilbert, 2002). The main concern is that over-reliance on CAQDAS might lead to over-use of coding, and thus loss of perspective, for example, by taking extracts or quotes out of context (King, 2008). In addition, there is also a concern that over-reliance on CAQDAS might lead researchers to focus merely on procedural criteria (e.g. providing an audit trail and checklists as a safeguard) without due consideration of the underlying philosophical and methodological criteria (Salmona and Kaczynski, 2016). For example, Kelle (1997) argued that many researchers claim to be adopting a grounded-theory methodology for their studies when, in fact, they are just applying a "coding paradigm" using CAQDAS. Thus, it is important for researchers to focus on providing a transparent account of both their analytical and methodological processes, so as to enhance the quality and credibility of their research work. As recommended by King (2008, p. 141):

> …it is important to remember that CAQDAS is not an end in itself and if users become overly focused on the software itself in a misguided attempt to find a 'right' way of analysing data, it is likely that they will become entrenched in the detail and potentially reduce the ability to build theory. As building theory is the ultimate goal of most qualitative research, we need to ensure that we continue to operate at a deeper level and consider the methodological and philosophical issues relating to the research design.

The decision about whether to use CAQDAS or not, ultimately, is that of the researcher. It is important that researchers recognise the value of both manual and computer-assisted methods in qualitative data analysis (Welsh, 2002). Accordingly, it is suggested that researchers should remain open to and make use of the advantages of each. The University of Northampton (2017) provided a basic guide that could help researchers to decide whether they should use an application of CAQDAS (e.g. NVivo) or not for analysing their qualitative data (Table 6.3).

Table 6.3 A guide for helping researchers to choose between the use of CAQDAS or manual methods

Question	1. CAQDAS	2. Manual analysis
The amount of data you are collecting?	Big amount – more than 10–15* interviews/texts/ images, and so on	Small amount – fewer than 10* interviews/texts/ images, and so on
Length or size of your pieces of data (interviews/texts/images)?	Interviews last at least 45 minutes or an hour* each/text over five pages each/large complex images*	Interviews under 30 minutes/text under two to three pages each/ small simple images*
Collecting different types of data?	Yes – for example, video, images, audio, observation notes, and interviews	No – just interviews or texts
Richness or complexity of the collected data?	Each interview/text/image has volumes of key themes and interesting points to capture	Each interview/text/image has one or two key themes or points for analysis
Will your analysis require linking different parts of the data together?	Yes, different themes will probably come up under different questions/topics	No, each question/topic can be addressed separately

Source: Adapted from The University of Northampton (2017).
* The numbers provided within the table are rough estimates only.

How can NVivo support systematic literature reviews?

The use of NVivo, as a CAQDAS, can help the researcher to take the qualitative data analysis much further than it would have been possible manually in the time available. There are many textbooks that explain how to use NVivo for analysing qualitative data (see, for example, Richards, 1999; Bazeley and Richards, 2000; Jackson and Bazeley, 2019). However, much fewer studies have focused on the use of NVivo for literature reviews. From the experiences of the authors in conducting qualitative systematic literature reviews using NVivo, it is suggested that NVivo can help the researcher in various ways, which include mainly the following:

- Storing, organising, and managing a large dataset systematically. NVivo can make it possible for the researcher to access, organise, and manage large amounts of data easily, including published academic studies. Using NVivo, the researcher can also view which parts of the data have been coded, and retrieve them whenever necessary, at a glance. NVivo makes it possible for the researcher to create "thematic nodes", which act as coding containers (Figure 6.4). These nodes can then be organised and merged to create hierarchies using parent and child nodes, in this way supporting the development of themes, categories, and models. In addition, the use of NVivo can enable the researcher to explore the content of any node (i.e. coded references and the name of the source that was coded) whenever necessary, thereby allowing for an efficient review of coded data.
- Data linking. Using NVivo, the researcher can connect relevant data segments and documents to each other easily with hyperlinks. More specifically, NVivo makes it possible for the researcher to attach external files, links, and internal annotations to coded data to record referential information that might be significant to establish context. This data-linking capacity also makes it possible for the researcher to create more seamless links between literature review notes, coded data, and reflective commentaries and ideas recorded in memos.
- Carrying out a search electronically. The "Advanced Find" and the "Text Search" options within NVivo can help the researcher to locate keywords or segments of texts easily, leading to more reliable and accurate search results than doing it manually, simply because human error is ruled out. The search tool can be useful also in terms of enabling the researcher to interrogate and gain an overall impression of the data whenever required. For instance, while conducting "constant data comparison analysis" (Glaser and Strauss, 1967; Strauss and Corbin, 1998), it can help the researcher to find relevant information and codes throughout a large dataset in a speedy manner.
- Using queries to explore and analyse coded data. NVivo makes it possible for the researcher to conduct various queries, such as the: (i) Text Search Query; (ii) Word Frequency Query; (iii) Coding Query; and (iv) Matrix Coding Query (QSR International, 2014). The "Text Search Query" can be useful when conducting a content analysis, as it allows the researcher to search for and quantify the occurrences of a specific word, phrase, or concept, and

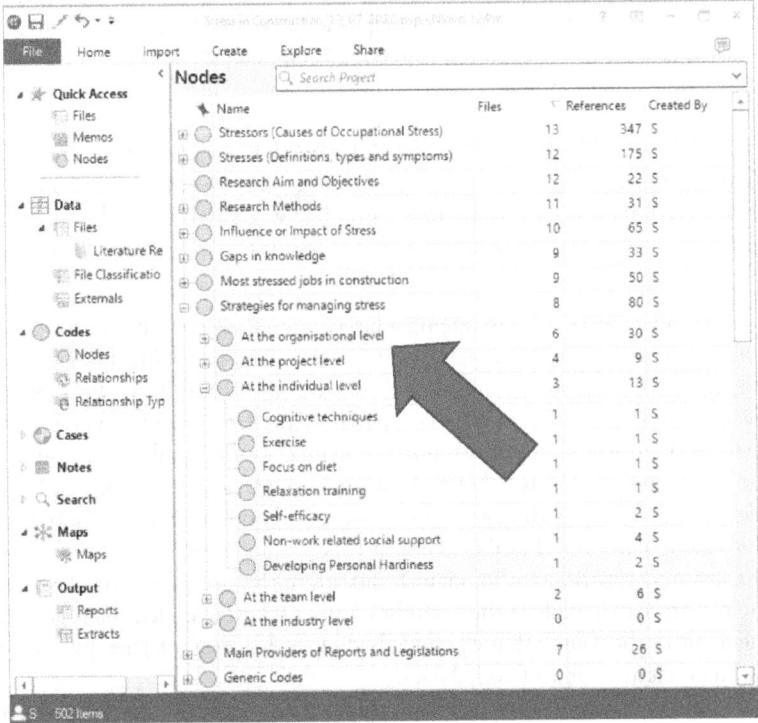

Figure 6.4 Nodes as "Code Containers" in NVivo, illustrating strategies for managing occupational stress at an individual level.

Source: Original.

to identify its references. The text search query can be useful also for a study in which a thematic analysis is used, as it allows the researcher to see where specific terms occur in the content of the project in the form of a Word Tree. Similarly, the "Word Frequency Query" allows the researcher to identify frequently occurring words or concepts in the content. This can be used, for example, to create Word Clouds based on the top 100 most frequently occurring words in the reviewed literature. The "Coding Query" allows the researcher to search for content based on how it is coded (e.g. to find content coded in two different code containers, i.e. nodes). The "Matrix Coding Query" makes it possible for the researcher to find coding intersections between two lists of items. This can be useful in enabling the researcher to interrogate the data, ask questions, and to seek explanations based on the resulting patterns in the data. The "Matrix Coding Query" can, for example, be used when conducting a thematic analysis to support the researcher in establishing relationships between emerging themes.

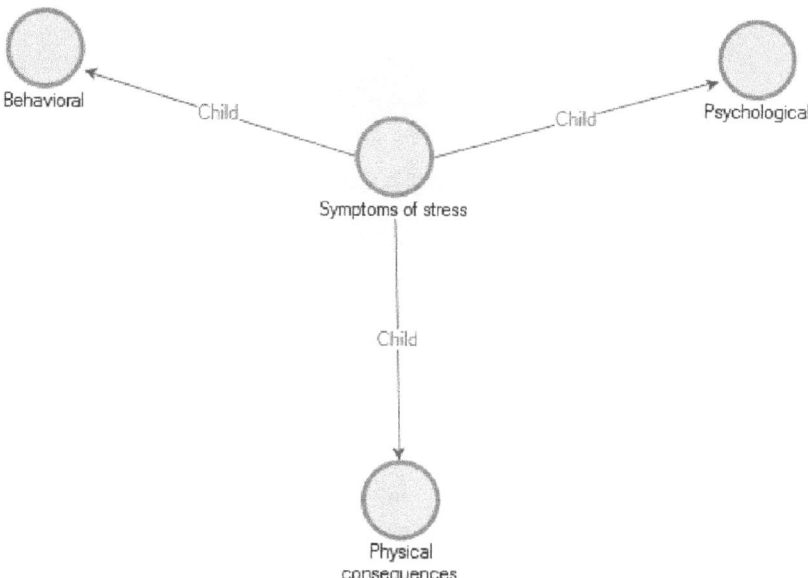

Figure 6.5 An example of an NVivo "Explore Diagram", illustrating a classification of the symptoms of occupational stress in construction.
Source: Original.

- Generating diagrams to visualise connections in the data. This is a useful option, which makes it possible for the researcher to generate an Explore Diagram that illustrates the linear relationships that exist within the node's folder, as shown in Figure 6.5. This option allows the researcher to visualise all project items connected to the selected node and to explore related items. The capacity to visualise linear connections in the data can be useful in making it possible for the researcher to explore and reflect on what is being coded.
- Creating charts or maps. This option is significant because it can help researchers to explore connections in the data, visualise patterned relationships between themes developed in the NVivo project, and present the findings and conclusions of their NVivo analyses in the form of mind, concept, or project maps. Figure 6.6, as an example, presents NVivo-12 outputs, based on the preliminary findings of an ongoing, funded research study that aims to investigate the root causes of occupational stress in the UK construction industry (Sarhan and Pretlove, 2021).
- Sharing the research project using NVivo Server and managing secure backups in multiple locations. The sharing option is significant because it allows researchers to share their work with others. So, for example, it could enable supervisor(s) to gain immediate access to the work whenever

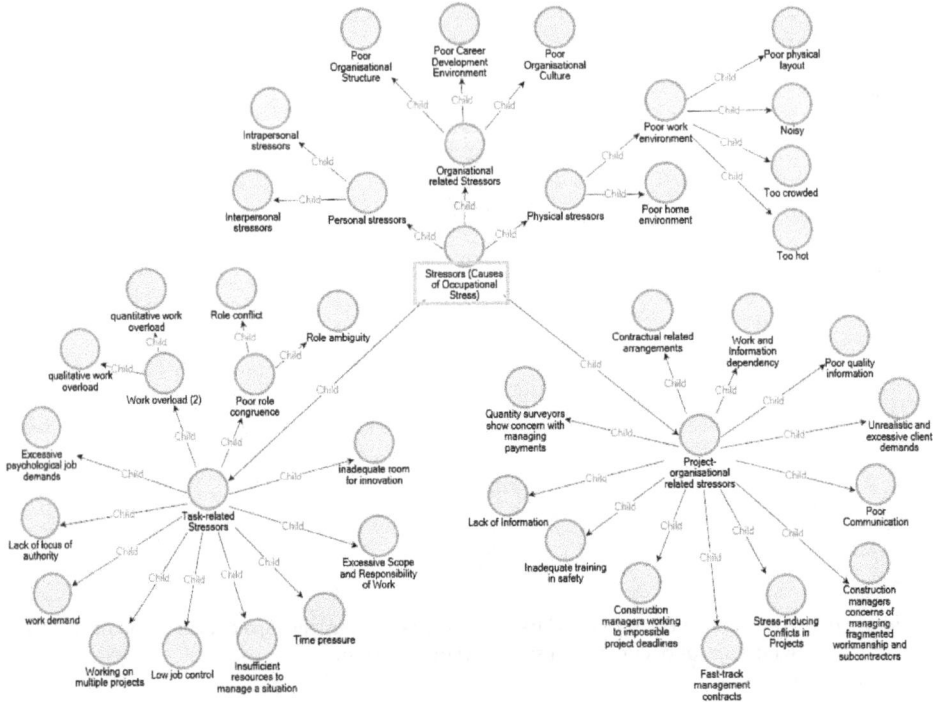

Figure 6.6 An example of an NVivo project map illustrating a detailed overview of sources of stress in construction.

Source: Original.

needed by the post-graduate student or researcher. The use of NVivo also reduces the researcher's worries about any losses or damages to the research project, as the software allows for saving and storing backups in multiple locations.

Conclusion

In this chapter, the role of traditional or narrative literature reviews in research have been discussed in relation to established secondary research methods that utilise published academic literature as a secondary data source for analysis. Comprehensive, methodical guidance for analysing published academic literature as secondary data when conducting systematic literature reviews has also been provided, with a focus on the use of a qualitative data analysis software application (NVivo) to support this process. This systematic approach can help students, ECRs, and academics who have considerable administrative duties, to enhance their chances of producing high-quality publications that report findings which have been synthesised from existing literature.

It is important to recognise that conducting systematic literature reviews does not mean just being comprehensive in searching for all relevant data, using explicit inclusion/exclusion criteria or using NVivo to analyse the collected data. Since theory building is often the main goal of qualitative research, it is imperative that researches follow a coherent and methodical approach throughout the entire research project, maintain a logic of the decisions made in the process, and provide evidence transparently. One of the significant aspects of any research design is the approach to reasoning in which it is incorporated. Therefore, it is necessary to understand how the reasoning and logical stances that are chosen can influence the approaches to data coding and analysis. CAQDAS application such as NVivo, if used appropriately, can be a tool that helps researchers to uncover connections in the data that would not be possible when performed manually. Use of NVivo or any other CAQDAS application can also help to improve the transparency, efficiency, outcomes, and validity of the research study. Although the approaches to data analysis have been discussed in this chapter using a systematic literature review, it should be noted that these approaches can be applied to other qualitative secondary research methods.

References

Ali, H. and Birley, S. (1999). Integrating deductive and inductive approaches in a study of new ventures and customer perceived risk. *Qualitative Market Research: An International Journal*, 2(2), pp. 103–110.

Badger, D., Nursten, J., Williams, P. and Woodward, M. (2000). Should all literature reviews be systematic? *Evaluation & Research in Education*, 14(3–4), pp. 220–230.

Bazeley, P. and Richards, L. (2000). *The NVivo Qualitative Project Book*, Sage, London.

Berelson, B. (1952). *Content Analysis in Communication Research*, The Free Press, Glencoe, IL.

Blair, E. (2015). A reflexive exploration of two qualitative data coding techniques. *Journal of Methods and Measurement in the Social Sciences*, 6(1), pp. 14–29.

Braun, V. and Clarke, V. (2006). Using thematic analysis in psychology. *Qualitative Research in Psychology*, 3(2), pp. 77–101.

Braun, V. and Clarke, V. (2012). Thematic analysis. In: H. Cooper, P. M. Camic, D. L. Long, A. T. Panter, D. Rindskopf and K. J. Sher (eds). *APA handbooks in psychology®. APA handbook of research methods in psychology, Vol. 2. Research designs: Quantitative, qualitative, neuropsychological, and biological.* American Psychological Association, pp. 57–71, doi: 10.1037/13620-004

Brewer, G. and Strahorn, S. (2012). Trust and the project management body of knowledge. *Engineering, Construction and Architectural Management*, 19(3), pp. 286–305.

Bringer, J., Johnston, L. and Brackenridge, C. (2004). Maximizing transparency in a doctoral thesis: The complexities of writing about the use of QSR*NVIVO within a grounded theory study. *Qualitative Research*, 4(2), pp. 247–265.

Charmaz, K. (2008). Grounded theory as an emergent method. In: S.N. Hesse-Biber and P. Leavy (eds.), *Handbook of Emergent Methods*, The Guildford Press, New York, pp. 155–172.

Creswell, J.W. (2007). *Qualitative inquiry and research design: Choosing among five approaches* (2nd ed.), Sage Publications, Inc., Thousand Oaks, CA.

Cronin, P., Ryan, F. and Coughlan, M. (2008). Undertaking a literature review: A step-by-step approach. *British Journal of Nursing*, 17(1), pp. 38–43.

Elo, S. and Kyngas, H. (2008). The qualitative content analysis process. *Journal of Advanced Nursing*, 62(1), pp. 107–115.

Gilbert, L.S. (2002). Going the distance: Closeness in qualitative data analysis software. *International Journal of Social Research Methodology*, 5(3), pp. 215–228.

Glaser, B. and Strauss, A. (1967). *The Discovery of Grounded Theory*, Aldine, New York.

Graneheim, U.H., Lindgrena, B. and Lundmana, B. (2017). Methodological challenges in qualitative content analysis: A discussion paper. *Nurse Education Today*, 56, pp. 29–34.

Grant, M. and Booth, A. (2009). A typology of reviews: An analysis of 14 review types and associated methodologies. *Health Information and Libraries Journal*, 26, pp. 91–108.

Guest, G., MacQueen, M.K. and Namey, E. (2012). *Applied Thematic Analysis*, Sage, London.

Hart, C. (1998). *Doing a Literature Review: Releasing the Social Science Research Imagination*, Sage, London.

Hsieh, H. and Shannon, S. (2005). Three approaches to qualitative content analysis. *Qualitative Health Research*, 15(9), pp. 277–1288.

Jackson, K. and Bazeley, P. (2019). *Qualitative Data Analysis with NVivo*, 3rd edition, Sage, London.

Kelle, U. (1997). Theory building in qualitative research and computer programmes for the management of textual data. *Sociological Research Online*, 2(2). Available at: http://socresonline.org.uk/2/2/1.html.

King, A. (2008). Using software to analyse qualitative data. In: A. Knight and L. Ruddock (eds.), *Advanced Research Methods for the Built Environment*, Blackwell, Oxford, pp. 1–13.

Neuendorf, K.A. (2019). Content analysis and thematic analysis. In: P. Brough (ed.), *Research Methods for Applied Psychologists: Design, Analysis and Reporting*, Routledge, New York, pp. 211–223.

Mzyece, D., Ndekugri, I.E. and Ankrah, N.A. (2019) Building Information Modelling (BIM) and the CDM regulations interoperability framework. *Engineering, Construction and Architectural Management*, 26(11), pp. 2682–2704.

Nowell, L., Norris, J., White, D. and Moules, N. (2017). Thematic analysis: Striving to meet the trustworthiness criteria. *International Journal of Qualitative Methods*, 16, pp. 1–13.

Pautasso, M. (2013). Ten simple rules for writing a literature review. *PLOS Computational Biology*, 9(7), pp. 1–4.

Pickering, C. and Byrne, J. (2014). The benefits of publishing systematic quantitative literature reviews for PhD candidates and other early-career researchers. *Higher Education Research & Development*, 33(3), pp. 534–548.

QSR International (2014). NVivo 10 for Windows – Getting Started. Available at: https://download.qsrinternational.com/Document/NVivo10/NVivo10-Getting-Started-Guide.pdf (Date of access: 11/11/2020).

Richards, L. (1999). *Using NVivo in Qualitative Research*, Sage, London.

Salmona, M. and Kaczynski, D. (2016). Don't blame the software: Using qualitative data analysis software successfully in doctoral research. *Forum: Qualitative Social Research*, 17(3), Art. 11.

Sarhan, S., Pasquire, C., Elnokaly, A. and Pretlove, S. (2019). Lean and sustainable construction: A systematic critical review of 25 years of IGLC Research. *Lean Construction Journal*, (2019), pp. 01–20. Available at: https://www.leanconstruction.org/learning/publications/lean-construction-journal/lcj-back-issues/2019-issue/

Sarhan, S. and Pretlove, S. (2021). Developing construction supply-chain management standards (CSCMS) for improving occupational stress management and productivity in construction projects. Unpublished report of an ongoing research project funded by B&CE's Charitable Trust. Available at: https://bandce.co.uk/corporate-responsibility/the-charitable-trust/grants-for-health-and-safety/occupational-health-research-award/

Sucharew, H. and Maurizio Macaluso, M. (2019). Methods for research evidence synthesis: The scoping review approach. *Journal of Hospital Medicine*, 7, pp. 416–418. doi:10.12788/jhm.3248.

Siddaway, A., Wood, M. and Hedges, V. (2019). How to do a systematic review: A best practice guide to conducting and reporting narrative reviews, meta-analyses, and meta-syntheses. *Annual Review of Psychology*, 70, pp. 747–770.

Strauss, A. and Corbin, J. (1998). *Basics of Qualitative Research Techniques and Procedures for Developing Grounded Theory*, Sage Publications, Inc., Thousand Oaks, CA.

The University of Northampton. (2017). How does Nvivo work? Available at: https://nile.northampton.ac.uk/bbcswebdav/pid-1547923-dt-content-rid-2423050_1/courses/Centre-for-Achievement-and-Performance/Research/ [Accessed on: 11 April 2017].

Vaismoradi, M., Jones, J., Turunen, H. and Snelgrove, S. (2016). Theme development in qualitative content analysis and thematic analysis. *Journal of Nursing Education and Practice*, 6(5), pp. 100–110.

Vaismoradi, M. and Snelgrove, S. (2019). Theme in qualitative content analysis and thematic analysis. *Forum: Qualitative Social Research*, 20(3), Art. 23.

Vaismoradi, M., Turunen, H. and Bondas, T. (2013). Content analysis and thematic analysis: Implications for conducting a qualitative descriptive study. *Nursing and Health Sciences*, 15, pp. 398–405.

Welsh, E. (2002). Dealing with data: Using NVivo in the qualitative data analysis process. *Forum Qualitative Sozialforschung / Forum: Qualitative Social Research*, 3(2), Art. 26.

7 A systematic literature review evaluating sustainable energy growth in Qatar using the PICO model

Redouane Sarrakh, Suresh Renukappa, and Subashini Suresh

Introduction

The industrial economy transformed the lifestyle of the human being and improved the standards of living. Carbon-based economies had driven humanity to invest more in technology and innovation, making life easier and more efficient. However, humanity's development has had a downside too. The increase of the earth's core temperature, the increasing number of pollutant-based particles in the air, and the continuous threats to the marine world are only a few examples, amongst a great deal of human activities that have an impact on the environment (Sneddon *et al.*, 2006). Therefore, policy-makers are required to make huge efforts to contain these problems before they intensify and result in catastrophic consequences for the planet (Boyd, 2003).

Sustainable development could be seen as a reactive response to the unhealthy regime of human activities and how they are carried out. The concept had been receiving a great deal of attention, especially after the Brundtland Report in 1987, and since then has become the focus of decision-makers all over the world (Singh *et al.*, 2012). However, human development had become highly dependent on unsustainable forms of resources, such as coal, oil, and gas, and the shift towards an economy with low dependence on fossil fuel would not be an easy task. The abundant oil and gas reserves within the Gulf countries made the region one of the energy leaders worldwide, as they are located collectively over up to 48% of the world's oil reserves and produce more than 85% of the world's natural gas (EIA, 2017). However, the significant increase in domestic consumption of energy threatens the region's future energy prospect and its position as the world's top oil and gas exporter. Much like the rest of the Gulf countries, Qatar, a small state within the region, achieved important economic growth because of its extraordinary gas reserves, making it the world's leading exporter of natural gas. However, British Petroleum (BP, 2017) predicted that Australia would overtake Qatar as the largest exporter of liquefied natural gas (LNG) by 2019. Similar to several countries that rely heavily on fossil fuel resources for their growth, Qatar has to make several changes to its social, economic, political, energy, and environmental structure to accommodate the necessary changes to ensure sustainable development. Therefore, two inter-connected streams of studies are discussed in this chapter, identified as:

- Evaluation of Qatar's current sustainable development plan and strategy;
- Performance of sustainability strategies within the oil and gas industry.

These discussions are indicative of the literature review that still needs to be conducted on the background context of the subject of interest, and the research gap that must be addressed even before applying a systematic literature review methodology.

Theoretical background

The 1960s and 1970s were a period marked by the population's high concerns about environmental problems and their impact on humanity, and the number of debates relating to humankind activities to address these problems increased considerably (Turner, 1988). Therefore, questions regarding the effectiveness of the conventional growth objectives, strategies, and policies started to surface in public debates (Dresner, 2007). The term "sustainable development" was first introduced in 1969, when it was mentioned in an official document signed by 33 African countries (IUCN, 1980). It was during the 1980s that the main objective of the International Union for Conservation of Nature and Natural Resources (IUCN) became the achievement of sustainable development through the preservation of nature's resources (IUCN, 1980). However, the report was criticised for being limited and focusing only on ecological sustainability and not linking it to either society or economy. Therefore, the necessity for a clearer and more inclusive concept became essential. Seven years later, and during the United Nations Environment Programme, the World Commission on Environment and Development (WCED) published a report titled *Our Common Future*, which became known as the Brundtland Report (WCED, 1987) that gave sustainable development a broader and better understanding. Brundtland defined sustainability as "development that meets the needs of the present without compromising the ability of the future to meet their own needs" (WECS, 1987). The report introduced the concept of integrating environmental policies within developmental strategies, refuting the common beliefs that protecting the environment could be done only at the loss of economic development. However; the report was criticised for lacking a clear, theoretical base (Simon, 1989) and being vague, leaving room for interpretation. Lèlè (1991) reported that the concept as introduced remained elusive and could be used wrongly to solicit funds. Still, Brundtland's report was the first to link global goals of sustainable development to changes in politics and society successfully. For instance, the elimination of poverty, equal distribution of natural resources, new methods of ensuring population control, changes in lifestyles, appropriate technological development, and necessary institutional changes (WCED, 1987), thus giving great importance to inter-generational and intra-generational equity in the use of resources (Baker, 2016).

The amount of literature regarding sustainable development and sustainability within organisations has seen an important increase in the last decade, especially in business firms (Renukappa *et al.*, 2014). From an organisational perspective, sustainable development is a holistic approach that addresses social, economic,

and environmental issues that would be beneficial for current and future generations of the stakeholders concerned (Renukappa *et al.*, 2012). However, Brundtland's definition of sustainability is difficult to adapt at this level. Therefore,
several scholars argue that the concept of the triple bottom line, developed by
Elkington (1998), has the best fit for organisations to follow to implement sustainability practices within their activities. Elkington (1998) developed the triple
bottom line concept to be a more practical definition of sustainable development,
a concept according to which the economic (profit), social (people), and environmental (planet) performances of organisations can be assessed simultaneously
and equally. Elkington (1998) relates corporate progression not only to economic
growth, but also to environmental sustainability and social responsibility. Therefore, managing these three aspects simultaneously represents one of the biggest
challenges for organisations (Epstein *et al.*, 2010).

Research methodology

The methodology used in this chapter was based on the systematic literature review approach, which has been used in a wide range of studies, such as health
care, social sciences, and education (Boaz *et al.*, 2002). The systematic review
approach was developed to collect available data systematically, filter them according to the credibility of sources, analyse filtered data to determine its overall
effect, and finally, disseminate the data based on their effectiveness (Higgins and
Green, 2011). Several studies that adopted the systematic literature review approach have been well received amongst academics and industry practitioners,
especially in social sciences (Adams *et al.*, 2016).

The systematic literature review approach is different from the traditional narrative review, as it adopts a replicable, scientific, and transparent process with
the objective to minimise bias through extensive literature exploration of existing published and unpublished studies (Tranfield *et al.*, 2003). This approach was
adopted for this chapter because of:

- the great interest Qatar has received since the early 2000s from academics
  and research institutions, giving rise to a huge number of research studies
  on Qatar's sustainability strategies and the sustainability of its energy sector;
- the uncertainty of the effectiveness of the already implemented strategies in
  the country and its energy sector.

Systematic review steps

Although the systematic review approach is relatively new compared to the traditional literature review method, a methodological process (see Table 7.1) was
reported by Higgins and Green (2011).

For the study presented in this chapter, the systematic literature review approach was followed with qualitative analysis of the resources, to provide clarifications about Qatar's current sustainability strategies, both overall and in its oil

Table 7.1 Stages of systematic review

Stage I – Plan the review

Phase 0	Identify the need for a review
Phase 1	Prepare a proposal for a review
Phase 2	Develop a review protocol

Stage II – Conduct the review

Phase 3	Identify research
Phase 4	Select studies
Phase 5	Assess study quality
Phase 6	Extract data
Phase 7	Synthesise data

Stage III – Report and disseminate

Phase 8	Report and make recommendations
Phase 9	Put the data into practice
Phase 10	Keep the review up to date

Source: Original.

and gas sector. At the planning stage, a review panel was formed of a number of academics and industry practitioners with expertise in both research methodology and sustainable development, as proposed by Tranfield *et al.* (2003). The process of the first stage was carried out by the panel through regular meetings, where disputes regarding the inclusion or exclusion of studies were resolved. The material gained from this approach was treated as questions and issues which would be of interest to academics, industry practitioners, and policy-makers. The steps followed within this study were as follows:

- Identify keywords and terms, based on a scoping study, available literature, and the panel's suggestions.
- Identify the most appropriate search strings.
- Compile a full list of the information search output of all the articles and papers reviewed.
- Incorporate studies that meet the inclusion criteria, as specified, into the research.
- Ask multiple reviewers to consider the inclusion/exclusion of resources, which can be subjective, and resolve disagreements during the panel meetings.

Research question

Formulating a carefully identified and well-constructed research question guarantees a focus on the research scope to avoid unrelated searching and to ensure the review of only useful information (Akobeng, 2005). A poor or unidentified question has the risk of being time-consuming, as the research would turn out to be significantly large and non-systematic. Therefore, it is essential for the researcher to frame and clarify the research question carefully to ensure a successful application of systematic literature review. Petticrew and Roberts (2006) argued that breaking the review question into sub-questions ensures a better framing and formulation of the question. The population, intervention, control, and outcomes (PICO) model is a tool that could be applied in this case.

The PICO model is a concept introduced originally as part of the guidance to assist in standardising the formulation of clinical research questions in the medical field, so that a literature review is carried out to answer them. The model was later adapted in social sciences studies to encourage researchers to consider different components when formulating review questions. The elements of the PICO Model include the following:

- Population (P): What population is the researcher interested in studying? This includes a clear identification of the population to be studied to eliminate any possibility of ambiguity.
- Intervention (I): What intervention is the researcher interested in reviewing? This could be one or multiple interventions, depending on the researcher's approach and chosen population.
- Comparison (C): To what is the intervention being compared?
- Outcome (O): What outcome does the researcher hope for from the proposed intervention? It is crucial to identify which outcomes are the most relevant to the question, in order to ensure an efficient collection of information.

The PICO Model has been used by Stone (2002) (Table 7.2).

What are the key sustainability strategies that have been implemented in the Qatar oil and gas industry?

Reliability of inclusion/exclusion decisions

Three panels, composed of a mixture of three members each, including academics with engineering, policy, business, and energy backgrounds, industry practitioners from the energy sector, and researchers were formed, first, to set the inclusion

Table 7.2 Description of the PICO model

Acronym	Definition	Description
P	Population	Qatar oil and gas industry
I	Intervention	Government/organisation sustainability activities
C	Comparison	Qatar's situation before implementing sustainability
O	Outcome	The country and industry's sustainability performance

Source: Original.

standards and, second, to review the sources to decide whether to include them or not. The inclusion/exclusion criteria were applied to the full content of the reviewed studies. Each reviewer of the review panels analysed the proposed articles separately to assess the extent to which the focus of the articles was on topics related to Qatar or its oil and industry's sustainable development strategies. Then, coders resolved the disagreements through discussion. A similar process was undertaken by Garcia *et al.* (2015), when they studied the factors that influence entrepreneurship.

The review focused on peer-reviewed journal articles that were available in five different databases: Science Direct, ProQuest, Google Scholar, Dawsonera, and Scopus. The reports of the Government and international associations also were considered considering the nature of the study (Table 7.3).

Generation and analysis of keywords

The members of the panels identified the keywords of the research based on their previous experience, using brainstorming sessions during the panel meetings. A list of the main keywords and terms was drafted and grouped by the researchers of the study into different categories. The list comprised 17 keywords classified under

Table 7.3 Criteria for inclusion/exclusion of studies

	Inclusion criteria	Exclusion criteria
Date	2008–2017	Prior to 2008
Geographic location	Qatar	
Language	English	Papers not in English
Type	Original research paper and textbooks	Articles and book reviews, research notes, dissertations
Publications	Peer-reviewed articles, government reports, published textbooks, conference proceedings	Papers focusing on the technical areas
Participants	Organisations within the Qatar energy sector	
Design	Qualitative, quantitative, case study, survey, studies that used a validated methodology	Informal papers, with no research questions, no research process, and defined data
Focus	Does the study examine Qatar's current overall sustainability strategies? Does the study identify practices and strategies to embed sustainability within Qatar and its energy sector?	Studies with no relationship to sustainability

Source: Original.

Table 7.4 Target dates for preparation tasks

Duration (weeks)	Phase
3	Preparation and development of review protocol
7	Search for relevant studies
6	Inclusion assessment
6	Data extraction/collection
5	Data analysis/synthesis
5	Reporting and recommendation

Source: Original.

three major categories: sustainability policies, keywords related to the sector, and performance outcome. The keywords were: Qatar sustainable development; Oil and gas sector; Carbon-based policies; Sustainability performances; Qatar Vision 2030; Sustainability strategies; National development strategy; Qatar energy sustainability performance; Knowledge-based economy; Qatar environmental development; Qatar economic development; Qatar human development; Qatar social development; Ladder of sustainable development and Qatar; Qatar Ministry Of Energy and Industry; Sustainable development practices; GCC sustainability; and Sustainable development goals.

The search timeframe

Specific target dates were set at the start of the literature review process in order to finish the preparation tasks within the agreed upon schedule. The targets were set in collaboration with the authors and different panel members (Table 7.4).

Data extraction

Microsoft Excel spreadsheets, available on a shared, open access platform (Google Drive), were used as a tool for data extraction. All the reviewers put their extracted data on their specific sheet within the document to be analysed afterwards. This strategy was used because it provided an inexpensive and easy access solution and presented the data in a format that can be easily summarised and analysed. An example of the data extraction form is shown in Figure 7.1 and the research framework that was followed is illustrated in Figure 7.2.

Results and discussion

An initial literature review search resulted in a list of 1990 sources. The selected sources were screened according to the series of criteria mentioned in the previous section. The selection was narrowed to 82 sources. Several papers were excluded as they were not considered to be relevant to the research topic or did not meet the inclusion criteria. Among the selected studies, 41 were from peer-reviewed journal articles, 10 from governmental reports, 14 from corporate reports,

Reviewer:	Date:
Author:	Year:
Journal:	Database:

Years covered by the search:

Terms and keywords used:

Search strategy summary:

Research methodology:

Participants:

Emerging themes:

Synthesis:

Figure 7.1 Data extraction form for conducting the systematic literature review.

11 textbooks, and 4 conference proceedings. The findings were reported in two sections, namely energy sustainability strategies and Qatar energy sustainability performance.

Energy sustainability strategies

Qatar's abundance in oil and gas resources had a significant impact on Qatar's economy. However, this exceptional and rapid development had a downside as well, evident in the detrimental impact on the environment. Furthermore, further excessive consumption of fossil fuel would increase the risk of the long-term impact of climate change. Therefore, the Qatari Government placed the inter-generational fairness of non-renewable energy usage at the heart of its sustainable development plan for energy (General Secretariat for Planning and Development, 2009).

Following the steps of the Qatar National Vision (QNV) 2030, the Ministry of Energy and Industry began to implement its sustainability programme: Qatar Energy and Industry Sustainability Strategy (QEISS). The programme was implemented to demonstrate and support the sector's contribution to the country's sustainable development plans and strategies. Considering the sector's fast rate of development and diversity, the Ministry of Energy and Industry had embedded innovation and business excellence culture within the programme, to help guide companies within the sector to implement sustainability management within their operations. The ministry opted also to improve the sector's impact

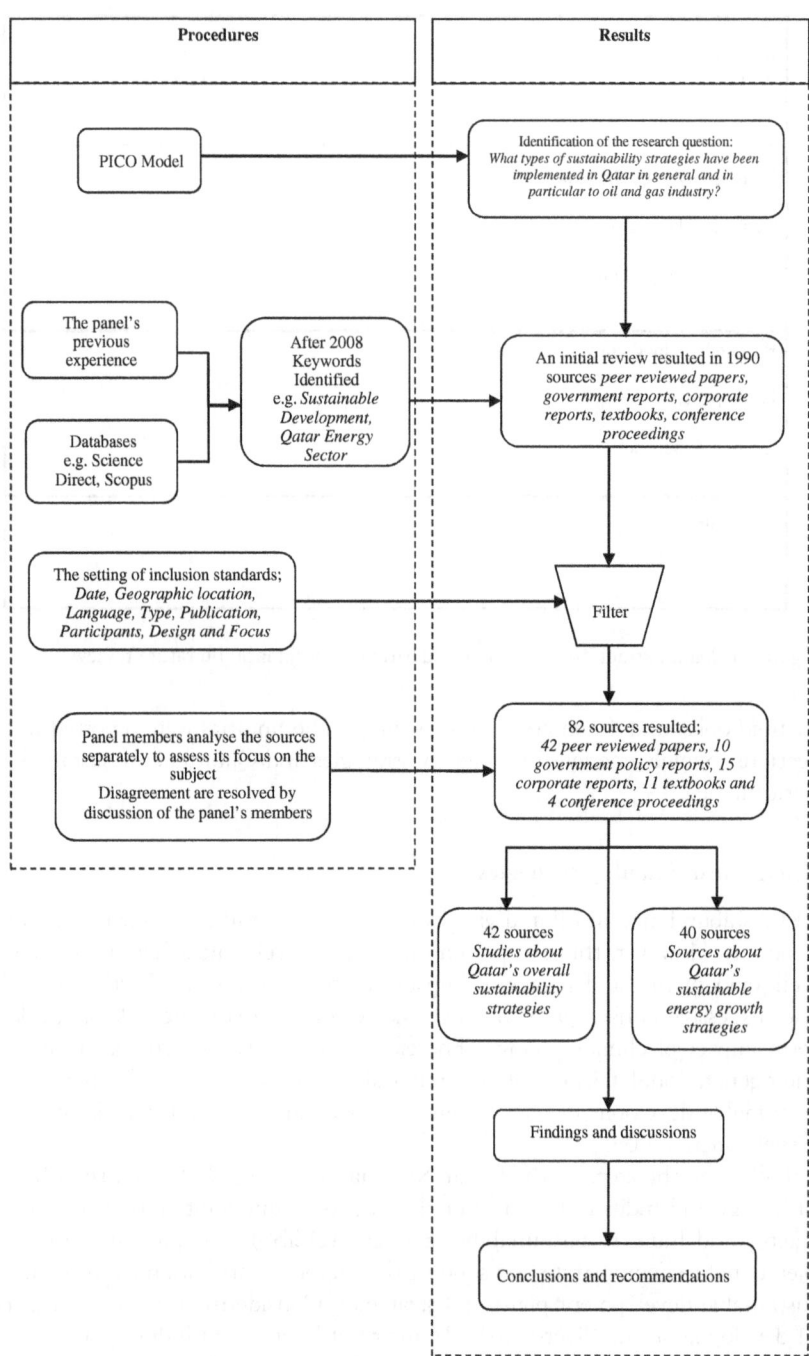

Figure 7.2 Research framework.

on the economy, environment, and society and to optimise its contribution to the country. The QEISS Programme is built around six components to ensure the integration and smooth implementation within the sector of policy input and formation, sustainability performance reporting, rewarding of good performance, sustainability performance assessment and targets, national and international engagement, and finally, sector sustainability strategy.

The sustainable development industry reporting was the first step taken by the ministry when establishing the QEISS. The reporting scheme was started initially as a voluntary programme to encourage companies to develop their own sustainable strategies and subsequently was made mandatory and incorporated other aspects to encourage existing and new companies to participate. Several sustainability aspects were prioritised in the sector's sustainability strategy and were categorised into six different areas: health and safety, environment, climate change, economic performance, society, and workforce. A wide range of indicators were used to measure the performance of companies in each area. The number of indicators increased each year since the start of the programme, which might represent a challenge to the participants as the overall number of data collected and provided would increase, thus posing questions about their quality.

The QEISS started with the participation of only 17 companies, four of which released their sustainability reports. The number increased to 36 participating companies by 2013, 21 of which released their sustainability reports, more than any country within the Gulf region. However, the number of companies participating was still small, especially considering the huge number of companies operating in the sector's eight sub-sectors (Qatar Online Directory, 2015). As the programme matured, amendments have been made by the ministry and participating companies on ways to improve the sector's sustainable development strategy. Thus, great importance was given to collected data and its quality, but the assurance level of companies' data was still a challenge, as only 55% of the data submitted in 2014 had been through some form of internal auditing process, approximately 27% was reviewed by an external auditor, and only 19% of the data were verified by international certifications such as the International Organization for Standardization or the British Standard for Occupational Health and Safety Assessment Series. This level of assurance was still low and did not meet the government's expectation and hope, thus raising questions about the data's level of transparency.

Qatar energy sustainability performance

Six areas were identified as the main components of sustainability performance within the Qatar energy sector. Each area contained several indicators that help companies to assess their sustainability status.

Workforce

The energy sector is one of the large employers of the national labour force together with the construction sector, owing to its frequent expansions and innovation projects. The country is trying to attract and retain the best international talents and invest simultaneously in its local population. The number of imported foreign labourers in the country increased to a point of seven to one in 2011 (Weber, 2011). Barrebi *et al.* (2009) explained that this derives from the structure of the Qatari labour market, because it diminishes any incentives for local Qataris to develop and acquire skills and knowledge required within the private sector. At the beginning of the last decade, the Ministry of Energy and Industry proposed a five-year strategic plan, referred to as "Qatarisation" to ensure 50% of the sector's employees are from a Qatar background (Qatar Ministry of Energy and Industry, 2014), but failed considering the desire of freshly graduated Qataris to work within the public sector as a result of the advantages and flexibility offered to Qatari citizens within the public sector (Randeree, 2012). Furthermore, based on the Qatar Statistics Authority (2014), the average salary per month within the public sector is QR22 781 (£4,786) and in the mixed public/private sector is QR18 700 (£3,740), while the average wage within the private sector is QR6883 (£1,446). Companies within the sector are still working towards the Qatarisation target by organising and participating in programmes and recruitment fairs for Qataris, such as the Qatar Independent Technical School Career Fair and Annual Career Fair for Qatar students in the United Kingdom. The companies also offer university students internships and partnerships. These efforts resulted 242 more Qataris being employed in the sector by 2014. Through sustainability strategies, the aim was to increase diversity within the sector to create an inclusive environment in the workforce, by employing people from different backgrounds and especially increasing the number of females in the workforce. Programmes included the Qatar International Business Women Forum led by Qatar Fuel Additives Company Limited (QAFAC), Ladies Group led by Qatar Petrochemical Company, and Maersk's mandatory inclusion of diversity e-learning for its employees. Programmes such as these led to an increase of 189 females being employed within the sector.

Health and safety

Companies within the energy sector had been part of four out of eight economic activities that contributed to non-fatal occupational injuries in 2013 (Qatar Ministry of Energy and Industry, 2014). This huge number of injuries had a negative impact on the sector in general and on companies, as it led to personal loss, reputational harm, fines, compensatory damages, and loss in production. This led to major institutions within the country – the Supreme Council of Health, the Ministry of Labour, the Ministry of Business and Trade and QP Ports, The Ministry of Environment and the Health, Safety, and Environment (HSE) regulations, and Enforcement Directorate – intervening to support the sector and its companies in managing their health and safety. By 2013, several laws and regulations were being implemented in the sector with regards to health and safety, where the HSE regulations and enforcement directorate was overseeing the activities. In the same year, the framework for Management of Major Accident Hazards was developed, which was completed in 2014. Upon completion, the framework became the basis for managing risk within the sector, with the objective to reduce major accident hazard risk to a level as low as reasonably practicable. The framework was intended also to prevent the occurrence of major accidents, minimise their consequences, and improve the sector's preparedness to react appropriately to potentially major emergencies. However, it is necessary for more audits and inspections to be implemented by companies to monitor their health and safety performances and check how rigorous they are, with the assistance of external auditors as they provide a more objective assurance of effectiveness and help identify more gaps and areas to improve.

Society

Qatar places great importance on its social development. The government's commitment has been stated clearly in the QNV 2030 and NDS by implementing systems and approaches dedicated to social welfare to ensure the provision of essential amenities and services to the community. In the same context, the Qatar Energy and Industry Sector plays an important role, as it is one of the main sources of financial revenue for the state, and its economic contributions finance national social development work and infrastructure in several areas such as education, health, sports, arts, and environment to the amount of $208 million (Qatar Ministry of Energy and Industry, 2014).

Energy companies have a great responsibility towards society. They should be the example of responsible corporates that contribute and act to society's benefit. Especially since corporate social responsibilities tend to be very limited to sporting events, education, health, and environment, while areas such as human rights, workers' rights, and anti-bribery and corruption measures are not covered by the energy companies' corporate social responsibilities (CSRs), mainly because there is a gap between the overall awareness and practices, considering that the industry is in its formative stage with regards to sustainable development (Kirat, 2015). Furthermore, it could be argued that organisations within the industry are selecting stakeholders with great importance given to media, public image, and pleasing the government above all. That is why the sector should be engaging more with the community to promote and support social development and working to protect the welfare of workers by implementing business integrity and ethical policies. A few energy companies have been engaged already in such CSR activities, like RasGas's commitment to having human rights audits for itself and its ten largest contractors led by an independent international expert on human rights management. Following the examples of RasGas and Qatar Shell, the QA-FAC ensured the inclusion of human rights criteria with all its contractors and suppliers, both national and international.

Environment

The sector as a whole, and companies as individual entities, started developing and adapting systems and strategies for environmental management, to mitigate the environmental impact of their activities and to generate short- and long-term benefits for the environment and people. Four aspects have been given great importance by the Qatari Government: water, waste, gas emissions, and spills. Enforcing environmental regulation is within the responsibilities of the Ministry of Environment, which has set specific requirements and environmental limits for companies that are in line with international standards.

Being one of the countries with the least amount of freshwater resources in the world, Qatar must manage its water resources strategically if it wants to realise its ambitions for long-term growth and sustainable development. Therefore, desalination of seawater, undertaken by QEWC, RLPC, RGPC and Q Power, remains the main source of freshwater used by households and companies. Several energy companies have on-site desalination and water treatment plants to process water used in their operations. The Ministry of Environment oversees companies' water management (and has placed limits on water withdrawal) and seawater discharge, and agreements were signed between companies and the ministry to regulate the quality and temperature of the water, depending on the companies' operations. The amount of water discharged by the energy sector experienced a 5% increase between 2013 and 2014, reaching more than 24 million cubic metres. This slight increase was caused by the expansion of the sector's production capacity, but the sector is working alongside the Ministry of Environment to reach a Zero Liquid Discharge to the sea.

The energy sector is responsible for emitting various pollutants that have an adverse effect on the air's quality, because components such as nitrogen oxides (NO_2), sulphur oxides (SO_2), and volatile organic compounds which, when combined with the high levels of dust in the country, can have a destructive impact on the environment and people's health. The country is faced with a serious challenge to find a way to maintain good air quality while continuing to increase industrial production. Therefore, the Ministry of Environment established acceptable emission levels for companies within the sector, based on their activities. The government is also providing specific guidance to companies regarding air quality management. Since the beginning of 2015, several companies have installed the Continuous Emissions Monitoring Systems within their production line, and two of them have set a five-year objective to control and reduce their NO_2 and SO_2 emissions by 75.5%.

Climate change

The Gulf region's extensive development of fossil fuel extraction and energy-based industry was responsible for the region's notable increase in greenhouse gas (GHG) emissions from 0.8% in 1973 to 5.2% of the worldwide emissions in 2012, with Qatar being responsible for 4.6% of the region's emissions (QatarGas, 2014) and the world's highest rate of GHG emissions per capita. The energy and industry sector's GHG emissions have increased by 2.2% between 2012 and 2013, owing to the production growth in several sub-sectors, with the LNG/NG companies being responsible for the largest share of emissions amounting to 47%.

Considering Qatar's geographic position, the country could suffer considerably from climate change. With the risks of rising sea levels, the country would be obliged to invest significantly in adaptation measures to protect the community and industry located in low-lying areas. The National Flaring and Venting Reduction Project is amongst the measures that the government has taken to reduce GHG emissions, caused mainly by flaring. Qatar has also established a national committee for climate change, which is responsible for formulating the country's climate policies. Moreover, during the Conference of the Parties (COP) 18 held in Doha in 2012, the Centre for Climate Research was launched in partnership with the Potsdam Institute for Climate Impact Research to encourage the exchange of climate change strategies amongst countries. The energy sector also has several initiatives in developing carbon capture and storage, because of the important role it plays in reducing GHG emissions. QAFAC has established the largest Carbon Dioxide Recovery Plant in the world (Qatar Ministry of Energy and Industry, 2014).

Economy

The energy sector contributes significantly to Qatar's economy, since it creates direct economic value by producing and exporting energy and other diverse products and services. The sector also provides an increasing number of well-paid jobs, thus helping to stimulate economic activities. The sector has suffered a minor challenge since 2014 when the stability of oil prices ended. After five years, oil that was traded for US$111 at the beginning of 2014 reached approximately US$27 at the start of 2016 (Beth, 2016), and oil and gas analysts and specialists came to the conclusion that the oil market had reached a turning point and would be flooded with a great excess in supply and shortage of demand (Khan, 2017). Despite these challenges, the energy sector was performing well, and its total hydro-carbon revenues continued to provide about 50% of the country's nominal GDP as a result of its variety of fossil fuel products and low production costs (Qatar Ministry of Energy and Industry, 2015). Furthermore, Qatar's strong fiscal reserves and savings rate played an important role in reducing the impact of the drop of oil prices and low energy demand on the country. The country also benefits from a low, fiscal, break-even point that protects it from any drop in oil prices. Moreover, Qatar is shifting towards short-term contracts with spot market sales instead of the long-term, oil-indexed contracts that remain linked to the volatility of oil prices (Wright, 2017).

Qatar has experienced an impressive growth in GDP since the early 1990s, out-performing its competitors in the Middle East (Alam *et al.*, 2016). In 2014, the country achieved a 6.5% increase in the GDP compared with 6.3% in 2013. This was mostly owing to the contribution of the construction and services sub-sectors, while the energy and industry sector was responsible for 40.4% of the country's GDP growth in 2014. In terms of revenues, the majority is generated by the LNG and oil and gas sub-sectors (56% and 26%, respectively), even though they experienced a drop in revenues in 2014. The contributions of the transport, storage, distribution, and refining sub-sectors' had increased by 11.9% and 9% between 2013 and 2014. Overall, the energy sub-sectors, apart from oil and gas and mining, have increased their revenues considerably since 2012.

Conclusion

"Sustainable development" is a term that embodies the integration of environmental, economic, and social aspects into one model to promote equity in accessing the planet's limited resources. In terms of this new model, great importance is placed on economic development while considering ecological protection, while offering flexibility in adoption within different cultures to meet their social needs.

The study presented in this chapter is representative of Qatar and its oil and gas industry, thus giving this research a largely exploratory nature, with its results being limited to Qatar, and of limited value in terms of generalisability. In this chapter, a case example of the systematic literature review has been presented that was based on the PICO model. The systematic literature review is limited to papers in the English language, and governmental and corporate reports, because of the availability of data. However, it should be acknowledged that with the focus of the study being on Qatar, which is an Arabic-speaking country, the research might be missing a number of studies that could be of importance. Therefore, the inclusion of Arabic papers might be considered in future studies.

Qatar's alarming energy consumption could endanger the country's long-term oil and gas reserves seriously and could hinder its economy directly. In accordance with the QNV 2030, the energy and industry sector introduced its sustainability program (QEISS). The sector identified six main areas of interest in which to evaluate the performance of the sector and its companies: workforce, health and safety, society, environment, climate change, and economy. In all of these areas an improvement in their sustainability performances has been experienced, except in the environment and specifically indicators related to air emissions, as the volume of GHG has increased since 2012 as a result of the sector's continuous growth in production. The programme was welcomed by the government that encouraged its development to ease and smooth its implementation within the sector and its companies. However, one of the main concerns regarding the strategy was the low number of participants within the sector. Although the overall number of participants had increased since its introduction, the number was still small and did not fulfil the vision of the country or the sector.

The QEISS Programme clearly provided guidance in measurement and management to the sector and companies with regards to sustainability, and the commitment of the ministry to continuous improvement of the programme was shown through its projects in different areas. However, it is worth mentioning that the last sustainability report by the sector, published in 2015, was for the year 2014. The sector had been focusing closely and working hard on lowering its impact on the environment and on improving accessibility of energy, but has been neglecting two of the energy pillars of sustainability – the renewability of energy resources and efficiency in energy conversion, distribution, and consumption. Even with the efforts of some companies to implement energy efficiency measures to decrease the consumption of energy, the overall energy consumption increased between 2013 and 2014. Therefore, companies within the sector should improve their performances relating to energy efficiency and reduce their consumption of fossil fuel resources by reviewing their business models to remain competitive and meet shareholders' expectations, especially in view of the rise of the Australian and American LNG market. The adoption of smart solutions is becoming essential to cope with the constant changes within the sector. New technologies such as "Distributed ledger – Blockchain" and "Big data" would have an important impact on the country's exploration and production activities, especially with Qatar's oil wells status

References

Adams, R., Jeanrenaud, S., Bessant, J., Denyer, D. and Overy, P. (2016). Sustainability oriented innovation: A systematic review. *International Journal of Management Review*, 18(1), pp. 180–205.

Akobeng, A. (2005). Principles of evidence-based medicine. *Archives of Disease in Childhood*, 90(8), pp. 837–840.

Alam, Q., Alam, S. and Jamil, A. (2016). Oil demand and price elasticity of energy consumption in the GCC countries: A panel cointegration analysis. *Business and Economic Horizons*, 12(2), pp. 63–74.

Baker, S. (2016). *Sustainable Development*, 2nd edition, Routledge, London.

Barrebi, C., Martorell, F. and Tanner, J. (2009). Qatar's labor markets at a crucial crossroad. *The Middle East Journal*, 63(3), pp. 421–442.

Beth, G. (2016). Mintec crude oil prices. Available at: https://www.thegrocer.co.uk/attachment?storycode=544337&attype=T&atcode=105679 [Accessed on: 12 November 2020].

Boaz, A., Ashby, D. and Young, K. (2002). *Systematic Reviews: What have they got to offer evidence based policy and practice?*. ESRC UK Centre for Evidence Based Policy and Practice, London.

Boyd, D. (2003). *Unnatural Law: Rethinking Canadian Environmental Law and Policy*, UBC Press, Vancouver.

British Petroleum (BP). (2017). *BP Energy Outlook 2035*, BP, London.

Dresner, S. (2007). *The Principles of Sustainability*, 2nd edition, Earthscan, London.

EIA. (2017). Qatar: International energy data and analysis (online). Available at: https://www.eia.gov/beta/international/analysis_includes/countries_long/Qatar/qatar.pdf [Accessed on: 6 July 2017].

Elkington, J. (1998). *Cannibals with Forks: The Triple Bottom Line of the 21st Century*, Capstone Publishing, Oxford.

Epstein, M., Buhovac, A. and Yuthas, K. (2010). Implementing sustainability: The role of leadership and organisational culture. *Strategic Finance*, 91(10), pp. 41–47.

Garcia, J., Puente, E. and Mazagatos, V. (2015). How affect relates to entrepreneurship: A systematic review of the literature and research agenda. *International Journal of Management Reviews*, 17(2), pp. 191–211.

General Secretariat for Planning and Development (GSPD). (2009). Advancing Sustainable Development: *Qatar National Vision 2030*, General Secretariat for Development Planning, Doha.

Higgins, J. and Green, S. (2011). *Cochrane Handbook for Systematic Reviews of Interventions*, Version 5.1.0, John Wiley & Sons, Ltd, Chichester.

IUCN. (1980). *World Conservation Strategy: Living Resource Conservation for Sustainable Development*, International Union for Conservation of Nature and Nature Resources, Switzerland.

Khan, M. (2017). Falling oil prices: Causes, consequences and policy implications. *Journal of Petroleum Science and Engineering*, 149(2), pp. 409–427.

Kirat, M. (2015). Corporate social responsibility in the oil and gas industry in Qatar: Perception and practices. *Public Relations Review*, 41(4), pp. 438–446.

Lèlè, M. (1991). Sustainable development: A critical review. *World Development*, 19(6), pp. 607–621.

Petticrew, M. and Roberts, H. (2006). *Systematic reviews in the social sciences*. Blackwell Publishing, Malden, MA.

Qatar Ministry of Energy and Industry. (2014). *Sustainability in the Qatar Energy and Industry Sector: Contributing to Qatar's Sustainable Development*, Ministry of Energy and Industry, Doha.

Qatar Ministry of Energy and Industry. (2015). *Sustainability in the Qatar Energy and Industry Sector: Contributing to Qatar's Sustainable Development*, Ministry of Energy and Industry, Doha.

Qatar Online Directory. (2015). Companies in Doha-Qatar (Online). Available at: http://www.qataronlinedirectory.com/150221/company-list/companies-qatar/ [Accessed on: 13 July 2018].

QatarGas. (2014). *Sustainability Report 2013*, QatarGas, Doha.

Randeree, K. (2012). *Workforce Nationalization in the Gulf Cooperation Council States*, Center for International and Regional Studies, Georgetown University School of Foreign Service, Qatar.

Renukappa, S., Egbu, C., Akintoye, A. and Goulding, J. (2012). A critical reflection on sustainability within the UK industrial sectors. *Construction Innovations: Information, Process, Management*, 12(3), pp. 317–334.

Renukappa, S., Egbu, C., Akintoye, A. and Suresh, S. (2014). Drivers for embedding sustainability initiatives within selected UK industrial sectors. *Journal of International REAL Estate and Construction Studies*, 3(1), pp. 1–22.

Simon, D. (1989). Sustainable development: Theoretical construct or attainable goal? *Environmental Conservation*, 16(1), pp. 41–48.

Singh, R., Murty, H., Gupta, S. and Dikshit, A. (2012). An overview of sustainable assessment methodologies. *Ecological Indicators*, 15(2), pp. 281–299.

Sneddon, C., Howarth, R. and Norgaard, R. (2006). Sustainable development in a post-Brundtland world. *Ecological Economics*, 57(2), pp. 253–268.

Stone, P. (2002). Popping the PICO question in research evidence based practice. *Applied Nurse Research*, 15(3), pp. 197–198.

Tranfield, D., Denyer, D. and Smart, P. (2003). Towards a methodology for developing evidence-informed management knowledge by means of systematic review. *British Journal Management*, 14(1), pp. 207–233.

Turner, R. (1988). *Sustainable Environmental Management: Principles and Practice*, Belhaven, London.

Weber, A. (2011). Youth unemployment and sustainable development: Case study of Qatar. *Revista de asistenta sociata*, 13(1), pp. 47–57.

World Commission on Environment and Development (WCED). (1987). *Our Common Future*, United Nations, Oxford University Press, Oxford, UK

Wright, S. (2017). Qatar's LNG: Impact of the changing East-Asian market. *Middle East Policy*, 24(1), pp. 154–165.

8 Understanding legal research in the built environment

Joseph Mante

Introduction

The built environment sector covers a wide spectrum of activities. These include design, procurement, construction, and management of structures within human settlements. The success of these activities depends partly on the nature of the contractual relationships between clients, designers, contractors, and the various professionals on projects and parties within the supply chain. Then, there is compliance that is required with relevant statutory and regulatory frameworks. The principles applicable in this case are largely normative and are concerned with value judgements based on, and resulting in, the making of valid prescriptive propositions. It has been argued that the tools of social science research, based on empirical investigations, are ill-equipped to determine true knowledge in law. Consequently, in doctrinal legal research, methods are adopted of which the aim is to identify and explain what the law is by logic, analogy, and deduction. The sources for such an exercise include legislation and case law. Doctrinal legal research might entail reliance also on secondary sources such as previous literature and legal commentaries. Legal research, therefore, relies on both primary and secondary data sources. There are significant lines of enquiry in built environment research which can benefit legitimately from legal research, especially the doctrinal approach. These include enquiries relating to interpretation of contractual clauses, determination of breach, issues of compensation for breach, decisions relating to violation of statutory provisions, and so on. The built environment researcher who is familiar with legal research approaches will be richer for it.

The aim of this chapter is to provide an overview of legal research by setting out how the doctrinal legal methodology, in its basic form, can be used in the context of built environment research for primary and secondary research. This chapter begins with brief information about the nature of law and a legal system generally. This is followed by a very brief examination of the philosophical perspectives of legal research and the different types of research approaches, with emphasis on the doctrinal approach, which is the focus of this chapter. For the rest of this chapter, the basic processes of doctrinal legal research have been described and the key steps explained.

Nature and relevance of law

Modern societies require some degree of order to survive and thrive, and this is mainly achieved through the application of law. What then is law? This is a theoretical question that has been debated for centuries among lawyers. Different scholars have attempted to provide different definitions of law with limited successes. Some define law as universal, natural principles steeped in right reason (or, better still, common sense). Others see law as a human creation. Austin (1832) defined law as a command of a political superior backed by sanction; a body of rules created and enforced by a sovereign or a political superior. Like the proverbial description of an elephant, each of these explanations of law is factual in some respects but do not capture every aspect of the nature of law. For the purposes of this chapter, the definition of law by Glanville William, a pre-eminent 20th-century legal scholar, has been adopted. Glanville William noted that "Law is the cement of society and also an essential medium of change" (Smith and William, 2006).

Laws are rules and regulations which are used to govern individual behaviours, preserve orderly running of society, and are veritable tools for all forms of social engineering. Two parties (an employer and a contractor) intending to have a business relationship will require some arrangement by which they will spell out their rights and obligations. Here, contract law provides the framework within which the parties' behaviour can be regulated. All aspects of human settlements, including how they are built – their design, procurement, construction, and management – are governed by different aspects of law. Consequently, built environment researchers ought to be interested in how legal research is conducted.

Legal traditions and systems

Different societies think about law differently. The diverse ways in which different cultures think about law are referred to broadly as legal traditions. Some of the major legal traditions in the world are the Common law, Civil law, Socialist law, Islamic Law, and Customary law. The term "common law" is used in many senses. As a legal tradition, it refers to the approach to law – its creation, development, organisation, and application – by cultures which follow the Anglo-Saxon (English) way of thinking (Örücü, 2008). The laws of countries such as England, Australia, New Zealand, Nigeria, parts of Canada, and the United States reflect this tradition.

Similar to the term "common law", civil law also has different connotations. As a legal tradition, it refers to how people who follow the Roman-Germanic culture think about law. The laws of many countries in continental Europe, including France, Germany, and Italy, operate under the civil law tradition. The differences in approach to how law is viewed and applied among legal traditions are enormous. For instance, according to the common law tradition, the development of law is viewed mainly as an inductive process, whereas according to civil law tradition, the same process is viewed as being deductive in nature. Consequently, many legal principles under the common law tradition have come into existence through centuries of evolution and development from practices of local communities and groupings by a process of repeated application and refinement by courts. Comparatively, many laws in countries operating under the civil law tradition have been enacted by sovereigns or authorities, compiled into Codes and used as guiding principles for human transactions (Örücü, 2008). As a necessary consequence, the approaches to application of the law also differ from one legal tradition to another. Also note that some jurisdictions do possess features of one or two legal traditions mixed into a common system. These are often referred to as mixed or hybrid jurisdictions (Tetley, 1999). In this chapter, the focus is on the common law tradition.

The practice of law as rules or principles would achieve little if there were no structures or processes through which the rules and principles are administered. This is where the concept of a legal system becomes important. A legal system has been defined as entailing a set of legal *institutions, procedures*, and *rules* (Merryman, 1985). A legal system consists of a set of laws applicable to a person or particular jurisdiction, how these rules are administered (procedure or method) and by whom (structures or institutions). The English legal system, for instance, applies to England and Wales. It has a distinctive set of laws derived mainly from legislation and judicial decisions. The laws are administered through courts and quasi-judicial bodies. Individuals or entities seeking to enforce the law are required to follow certain procedures, depending on whether the type of enforcement is criminal or civil. Anyone intending to carry out legal research in England, for example, must learn about the laws of England and Wales, the court structure, and the procedure by which these courts operate. It is essential to understand that there are different types of laws applicable to different situations. Criminal laws relate to matters concerning public safety, security, and order. The civil law, such as contract and tort, is used to regulate inter-personal, inter-entity transactions and resulting disagreements, which do not involve the State directly.

Again, it is important to note that the concept of "hierarchy" is essential in English Law. Both laws and courts exist within a structure of hierarchy. This understanding is crucial, especially for researchers. The higher the law in terms of hierarchy, the more weight is attached to it. In this regard, legislation is superior to case law. However, the relationship between the two sources of law might be more symbiotic than is often admitted. Judicial interpretation of legislation might carry more weight in some instances as it may qualify what might appear to be the literal meaning of a provision of a piece of legislation. Further, the higher the court, the more weight is accorded to its decisions. The decision of the Supreme Court, the highest court in the United Kingdom, supersedes that of the Court of Appeal and every other court below it. Similarly, the decision of the Court of Appeal, the second highest court in England, overrides the decisions of all other courts below it on a similar matter. A researcher using case law must be aware of this as it has implications for what may qualify as law.

Legal research approaches

Traditionally, the focus of legal research has been on the nature and meaning of law and the methodology has been mainly doctrinal, relying heavily on textual analysis. That said, it is also true that the search for truth in law has also involved society's understanding and interaction with law. This kind of enquiry often has combined both doctrinal and empirical approaches (Davies, 2020). There are those who hold the view that law is a social construct and should be studied empirically and contextually (Banakar, 2009, Banakar, 2011). Freeman (2008) stated that since law is a system of norms and a form of social control based on certain patterns of human behaviour, both the normative aspect and the social dimension of law are legitimate fields of enquiry. In this chapter, however, the focus is on the normative inquiry, as it was assumed that readers of this book will be familiar already with the social science approaches to research.

Doctrinal legal research

The use of a doctrinal legal approach to research has dominated the field of law for centuries. Its intuitive application has been unquestioned until recently (Hutchinson and Duncan, 2012). It is a type of research which is focused on legal principles and concepts. An understanding of the concept of "doctrine" in law will make it easier for the reader to follow the nature and requirements of doctrinal legal research. The word has been defined as entailing "a synthesis of various rules, principles, norms, interpretative guidelines and values. It explains, makes coherent or justifies a segment of the law as part of a larger system of law..." (Mann, 2010, p. 197). Legal doctrine is central to the common law system of law. It plays a key role in the "development of the conceptual framework of the legal order and its legal methodology" and serves as an organising concept, describing and systematising the law (Van Hoecke and Warrington, 1998). Individual statutes and cases must fit into the existing legal framework. This essence of legal doctrine is aligned with the concept of judicial precedent. There must be a clear explanation of how new situations or decisions fit into the well-established paradigm. From the foregoing, it is argued that the focus of doctrinal legal research is not only on how the law came into existence, who it affects, the impact of the law on society, or indeed, even pragmatic issues considered on a daily basis by lawyers; the focus is on finding out what the law is (describing the law), the process by which the normative character of specific legal concepts or principles is affirmed (interpreting the law), and how specific legal principles and concepts fit the broader legal order (systematising the law) (Van Hoecke and Warrington, 1998). This understanding of the doctrinal research approach is in line with the definition proffered by the Pearce Committee (1987), which identified the following four key areas of interest of doctrinal legal researchers:

1   Systematic exposition of legal principles;
2   Analysis of relationships between legal principles/concepts;
3   Explanation of areas of difficulty; and
4   Predicting future developments and proffering recommendations.

Table 8.1 Stages of doctrinal legal research

Stages	Process
1	Identify research questions or legal issues
2	Exposition of law – identify and state the law
3	Systematise the law – analyse and apply the law
4	Proffer future predictions and recommendations

Source: Original.

Doctrinal legal research is not as simple and straightforward as it may seem. The skills for it are honed over a long period of training and practice. For novice researchers (in the law) from an inter-disciplinary field such as the built environment, it might be daunting. These researchers stand the risk of furtively applying familiar social science research approaches to this enterprise without paying much attention to the peculiarities of doctrinal legal research. In that case, they might look for some signposting of what is required.

Further, there is no formulaic way of conducting doctrinal legal research; it is often an iterative process involving repeated engagement with the different elements of the discipline. That said, there are steps or signposts in the legal research process with which built environment researchers will be familiar, including identifying questions for investigation, exploring previous research on the subject, and using that information to clarify the research questions. The substance and focus of this exercise may differ for a social science researcher on one hand, and a doctrinal legal researcher on the other. Table 8.1 sets out the key procedural elements of the doctrinal research process. Each component is stated and explained briefly with examples related to the built environment.

Identify research questions or legal issues

It is important for a researcher using the doctrinal legal approach to understand the source of possible questions for investigation. These questions might arise as a result of an occurrence or development in society generally (such as the Grenfell Tower fire disaster) or a dispute that might require a legal outcome. In the context of built environment research, such questions may take the following forms:

a   Seeking an understanding of the position of the law on a subject (e.g. what is the law governing liability of builders in relation to defects or health and safety?);
b   Examining how existing law extends to new developments in society (e.g. how are the emerging issues in building information modelling or smart contracting addressed in terms of the current law on property and intellectual property?);
c   Explaining how old/new concepts should be interpreted and applied, (e.g. the meaning of mutual trust and co-operation under NEC 3 or NEC 4);

d   Evaluating the relationship between old and new concepts (e.g. how are the concepts of mutual trust and co-operation related to the much more popular concept of good faith?); and
e   Contract interpretation – dealing with difficulties and seeming contradictions.

Whilst establishing the context and raising questions for investigation might be common to built environment and doctrinal researchers, the latter will be guided in the process by a totally different conceptual framework, i.e. the legal system. Similarly, where and how social science researchers and doctrinal researchers look for answers to the research questions will differ. The former might consider observations of natural phenomena or statements from people affected by the relevant issues but the latter will have their attention focused steadily on the existing legal framework, legal norms, and judicial precedents for answers. The data for doctrinal researchers will include a review of the relevant literature but also, and more importantly, a search for the relevant, applicable law. This is the next aspect of the process – legal exposition.

Exposition of law: identify and state the law

An exposition of the law involves, among other things, what Van Hoecke and Warrington (2008) refer to as "describing the law". The "law" here refers to the legal principles relevant to finding answers to the legal issues or research questions identified. The process of describing the law entails a literal description of the law, an interpretation of it, and the determination of its validity. For instance, if researchers want to know about the rules concerning safety at a construction site in the United Kingdom, they might ask, "Which legal rules provide this information?". They might discover that regulations have been passed pursuant to the Health and Safety at Work Act of 1974 – the Construction (Design and Management) Regulations of 2015, which addresses this question. To have a deeper understanding of the current state of the law, they might decide to refer to previous regulations and how the current position of the law has evolved. Researchers might choose to examine the motivations for changes which culminated in the current law. All these inquiries would constitute part of the process of describing the law. To establish the validity of the current regulations, researchers might ask whether the provisions of this law are consistent with the content of the parent legislation, that is, the Health and Safety at Work Act of 1974. If not, a court is likely to invalidate the offending provisions of the regulations at some point. The foregoing example confirms the argument of Van Hoecke and Warrington (2008) that "describing the law" necessarily entails interpreting and validating the law.

The process of legal interpretation is carried out within a certain legal framework. What is acceptable as law or material facts (or indeed, reality in law) is determined by and perceived with reference to the conceptual framework referred to as the legal system (Van Hoecke and Warrington, 2008). In this sense, what is considered to be a relevant fact is not determined by what the parties affected by the law think, but what the law (taking account of what needs to be established in order to satisfy a legal threshold) requires.

Exposition of the law is not merely an exercise in logical reasoning. It is based on data – primary and secondary. In the legal context, this data may be in the form of legislation and judicial decisions. This explains why an effective legal exposition is only possible if the researcher has reasonable knowledge of the sources of law. In the United Kingdom, sources of law include legislation, delegated legislation, and judicial decisions (cases). Legislation comprises laws passed by parliament. There are thousands, if not millions, of pieces of legislation in force in the United Kingdom covering a wide spectrum of subjects, including construction, health, and safety, the environment, procurement, and so on. Examples of domestic legislation include the Health and Safety at Work Act of 1974; the Human Rights Act of 1998; Housing Grants, Construction and Regeneration Act of 1996; and the Local Democracy, Economic Development and Construction Act of 2009. Then there is EU legislation which remains applicable, regardless of BREXIT. Delegated legislation refers to regulations and by-laws enacted pursuant to powers conferred on a minister by legislation. An example of delegated legislation is the Construction (Design and Management) Regulations of 2015. Both legislation and delegated legislation can be accessed in hard copies in libraries and online through dedicated databases. Again, special skill is required to find legislation in hard copies. Librarians can be of immense help with this, but ideally, researchers should be able to use the library catalogue to trace where the relevant resources are located in the library. Then, there is the question of where specifically in the hefty bound and loose-leafed books a particular legislation is located. Alternatively, built environment researchers in the United Kingdom can rely on the on-line database of UK legislation at the following address: <http://www.legislation. gov.uk/>. Similar databases for legislation might exist in other jurisdictions.

Cases are records of judicial decisions over the years. Case law thus refers to legal principles which have developed as a result of consistent exposition and application of the law. These principles, like legislation, run into thousands (if not millions) and apply to different subject matter in society. A typical example of case law is the "neighbour principle" in *Donoghue v Stevenson*. This principle, addressing what will constitute negligence, was popularised by the then United Kingdom House of Lords in a decision in 1932. Since then, it has been explained, refined, qualified, and expanded by other courts. Similar to legislation, cases are reported in volumes called Law Reports. The Law Reports contain "processed data" from primary records of cases heard by courts. Thus, in a sense, the Law Reports could be regarded as secondary data. These secondary data are the staple of many a doctrinal legal researcher.

Each case is assigned a citation making it possible for researchers, lawyers, and judges to identify and retrieve copies when required. There are different forms of citation. Each citation system encapsulates some vital information about a case. This includes the name of the parties (title of the case), the year in which the decision was made or reported, the court which made the decision, and the page where the decision can be found. For instance, the citation for *Donoghue v Stevenson* is [1932] A.C.562. This citation means the case was decided in 1932, reported in the law report known as the "Appeal Cases", and can be found on page 562 of that report.

Sometimes cases may be reported in multiple law reports. So, *Donoghue v Stevenson* is also reported in the "All England Report" with the citation [1932] All ER 1.

Modern reports use what is referred to as neutral citations. This means essentially that no reference is made to specific law reports. The citation mentions the date of the decision, which court made the decision, and a special case number. For instance, here is the citation of the Supreme Court case *Robinson v Chief Constable of West Yorkshire Police*: [2018] UKSC 4. This citation means this is the decision of the UK Supreme Court delivered in 2018 with a unique reference number 4. This case was heard by the Court of Appeal before it was taken to the UK Supreme Court. The citation of the Court of Appeal decision is [2014] EWCA Civ 15. This citation means the decision was made in 2014 by the Civil Division (Civ) of the Court of Appeal of England and Wales (EWCA). Many law reports are now easily accessible online through dedicated databases such as Westlaw, Lexis Library, and Bailii.

Navigating these sources of law, to find which laws or legal principles are in force or are applicable to a specific situation, requires special skills. In this regard, built environment researchers venturing into doctrinal research must acquire, at least, basic aspects of these skills, such as finding the law, reading it along with relevant commentaries and background materials, and identifying the hierarchy or status of the relevant law in relation to others (Finch and Fafinski, 2019). Much of what is covered in the preceding paragraphs is about finding the law.

Reading the law requires a different set of skills. It requires an understanding of how both legislation and cases are structured. A case may run into hundreds of pages. The ability to identify what is relevant in a case is thus a very useful skill. Although there might be slight differences in the structure of judicial decisions, most of them will follow a common structure (Table 8.2):

Table 8.2 Case structure

Case structure	Explanation
1 Title	Ordinarily the names of the parties who were in court. In some instances, the names of the parties are not used
2 Head notes	A summary of the facts and principles applied or established in the case
3 Material facts	Detailed description of established and controversial facts relevant to the dispute
4 Issues/points of law	The legal questions that the court must address if it is to arrive at a conclusion one way or the other
5 Statement of the law	A discussion of relevant legal principles, often in generic terms and at times analogically
6 Legal analysis and application (judgment)	Application of the law, as discussed, to the facts established (by evidence). Discussion of the facts and evidence and application of the general legal principles, as discussed, to the specific case before the court, leading to a judicial decision. This is where the *ratio decidendi* and the decision and orders of the court will be found

Source: Original.

The part of every judicial decision, which is relevant to the purposes of iden-tifying the law, is the *ratio decidendi* (the legal principle on which the decision of a court is based). When dealing with case law, one judicial decision may affirm, qualify, or build on another. Consequently, the search for a legal principle appli-cable to a research question or legal issue necessarily entails some form of analysis of one or more relevant cases and or legislation.

Systematise the law: analyse and apply the law

Stating the law that is relevant to addressing a research question does not imply that the relevant legal issue has been resolved. The researcher must demonstrate how, in practical terms, the relevant law applies to or addresses the research ques-tion. This is achieved through various forms of legal analysis. Chynoweth (2008) identifies four of these as being deductive, analogical reasoning, inductive reason-ing, and policy consideration.

Deductive reasoning in law is quite similar to what pertains in the social sciences. Here, a general principle is applied to a factual and specific situation when the facts satisfy all relevant conditions. To succeed in establishing general negligence against a contractor on site, the aggrieved party must establish that the contractor owed it a duty of care in law, which has been breached and that the said breach has caused injury to the claimant or harm to the claimant's property (Giliker, 2017). Assuming there is evidence that all the legal conditions are satisfied, deductive reasoning will require that the judge or researcher concludes that negligence is established. Nonetheless, in practice, more analysis would be required before such conclusions could be drawn. Consequently, deductive reasoning is only of limited assistance. A researcher or a judge faced with such a dispute would be likely to refer to previous decisions, preferably of a higher court, on similar facts and apply that decision to the case. This style of legal analysis – analogical reasoning – is by far the most common method of judicial reasoning (Chynoweth, 2008).

The decision and reasoning in a case by a higher court may be extrapolated to future cases. Eventually, the "single case" becomes generalised, crystallising into an established legal principle and even a doctrine. Again, this approach to legal analysis – inductive reasoning – is well known in scientific research. It is par-ticularly common in single case experiments where outcomes have, in many in-stances, been replicated and ultimately become accepted propositions (Flyvbjerg, 2006). There are also instances where a mix of the approaches to legal analysis described earlier is deployed by the courts or researchers. In cases concerning negligence, judges have employed the different forms of legal analysis – deductive, analogical, inductive, and policy considerations – to connect the facts in those cases to the relevant law. From the case of *Donoghue v Stevenson (1932)*, the Courts have employed analogical and deductive reasoning consistently, applying the *ratio decidendi* of that case to similar cases in fact. In 2018, the UK Supreme Court clarified the criteria for establishing duty of care in negligence cases, again relying on the different types of legal reasoning identified above. The Supreme Court decided in *Robinson v Chief Constable of West Yorkshire Police (2018)*, a case involving alleged police negligence that, in cases relating to an established category of liabilities, courts were to consider what had been decided previously and follow existing precedents (analogical reasoning). The Supreme Court held that, in novel situations, the use of a combination of different legal reasoning was warranted. Starting with analogical reasoning, some sort of deductive reasoning and policy considerations should enable the extension of established, existing rules to novel situations. The key is to maintain coherence. Doctrinal researchers follow a similar process to what has been described, challenging existing assumptions in the process and proposing new ones.

Future predictions and recommendations

Note that legal analysis does not involve just harmonising the application of specific legal principles across similar scenarios. Legal analysis also involves how new laws fit into existing categories and how existing legal principles are re-interpreted to take into account of recent or future developments in society. In some instances, the application of common law has been extremely efficient in developing frameworks to address novel and future scenarios. In other instances, a more radical approach to re-systematising the law might be required. A typical example would be how the existing concepts of contract law can be used to respond to the emerging issues in smart contracting. This will necessarily call for re-systematising of existing doctrines such as consideration and intention (meeting of minds). Van Hoecke and Warrington (2008, p. 525) suggested that the process of re-systematising can be undertaken by "a (re)interpretation of the differing legal rules, in the light of a coherent unity, on the basis of a number of basic concepts and principles" such as judicial precedent.

Conclusion

Doctrinal legal research, in its basic form, is about locating, describing, interpreting, and systematising legal principles and concepts, with reference to the legal system as a conceptual framework. The sources for this process are legislation, cases, legal commentaries, and other literature about law. Although some of these sources are classified in social science terms as "secondary data", the data employed in doctrinal legal research are a combination of primary and secondary sources. What is different about doctrinal research compared with broader social science research is that data, whether primary or secondary, are processed differently within a different conceptual framework. Similarly, the outcomes of legal research might not resemble those of typical quantitative or qualitative research, but what is certain is that the outcome of a well-conducted doctrinal research is robust, logical, and consistent with the norms of its conceptual framework. Conclusions drawn from the process of identifying and stating the law, analysing and applying the law, and making future predictions would often supply a clear and robust normative response to any research question. The analytical process in doctrinal research has stood the test of time, partly for these very reasons.

References

Austin, J. (1832). *The Province of Jurisprudence Determined*, Richard Taylor, London.

Banakar, R. (2009). Law through sociology's looking glass: Conflict and competition in sociological studies of law. In: A. Denis and D. Kalekin-Fishman (eds). *The ISA Handbook in Contemporary Sociology*. Sage, London, pp. 58–73.

Banakar, R. (2011). *The Sociology of Law*. Sociopedia.isa. doi:10.1177/205684601134.

Black, T.R. (1999). *Doing Quantitative Research in the Social Sciences: An Integrated Approach to Research Design, Measurement and Statistics*, Sage, London.

Chynoweth, P. (2008). Legal research. In: A. Knight and L. Ruddock (ed.), *Advanced Research Methods in the Built Environment*, Blackwell Publishing Ltd, Chichester, pp. 28–38.

Cotterrell, R. (1998). Why must legal ideas be interpreted sociologically? *Journal of Law and Society*, 25(2), pp. 171–192.

Davies, G. (2020). The relationship between empirical legal studies and doctrinal legal. *Erasmus Law Review*, 2. doi:10.5553/ELR.000141.

Finch, E. and Fafinski, S. (2019). *Legal Skills*, 7th edition, Oxford University Press, Oxford.

Flyvbjerg, B. (2006). Five misunderstandings about case-study research. *Qualitative Inquiry*, 12(2), p. 219.

Freeman, M.D.A. (2008). *Lloyd's Introduction to Jurisprudence*, 8th edition, Sweet & Maxwell, London.

Giliker, P. (2017). *Tort*, Sweet & Maxwell, London.

Hutchinson, T. and Duncan, N. (2012). Defining and describing what we do: Doctrinal legal research. *Deakin Law Review*, 17(1), pp. 83–119.

Mann, T. (ed.) (2010). *Australian Law Dictionary*, Oxford University Press, Oxford.

Merryman, H. (1985). *The Civil Law Tradition: An Introduction to the Legal Systems of Western Europe and Latin America*, Stanford University Press, Palo Alto, CA.

Nelken, D. (1998). Blind insights? The limits of a reflexive sociology of law. *Journal of Law and Society*, 25(3), pp. 407–426.

Örücü, E. (2008). What is a mixed legal system: Exclusion or expansion? *Electronic Journal of Comparative Law*, 12(1). Available at: http://www.ejcl.org.

Pearce, D.C., Campbell, E., 1932-2010 & Harding, D.E. and Commonwealth Tertiary Education Commission (Australia) (1987). Australian Law Schools: A discipline assessment for the commonwealth tertiary education commission. Australian Government Publishing Service, Canberra. Available at:

Silbey, S. and Sarat, A. (1988). Dispute processing in law and legal scholarship: From institutional critique to the reconstruction of the juridical subject. *Denver UL Review*, 66, p. 437.

Smith, A.T.H. and William, G. (2006). *Learning the Law*, Sweet & Maxwell, London.

Stake, R.E. (1995). *The Art of Case Study Research*, Sage, London.

Summers, R.S. (1971). The technique element in law. Cornell Law Faculty Publications Paper 1192. Available at: http://scholarship.law.cornell.edu/facpub/1192.

Tetley, W. (1999). Mixed jurisdictions: Common law vs civil law (codified and uncodified) (part I). *Uniform Law Review*, 4(3), pp. 591–618.

Van Hoecke, M. and Warrington, M. (2008). Legal cultures, legal paradigms and legal doctrine: Towards a new model for comparative law. *International and Comparative Law Quarterly*, 47, p. 495.

9 Applying science mapping in built environment research

Amos Darko and Albert Ping-Chuen Chan

Introduction

Research in the built environment (BE) encompasses behavioural, affective, and cognitive components, and a systematic process of investigation that increases knowledge by answering an unanswered question or by solving an unsolved problem (Amaratunga *et al.*, 2002). BE research is based on two inquiry paradigms: positivism, which requires quantitative/experimental methods to test hypotheses; and interpretivism, which requires qualitative/naturalistic methods to produce hypotheses. While Amaratunga *et al.* (2002) discuss research in BE in terms of the two inquiry paradigms in detail, these paradigms have led to two key research types in BE: quantitative research and qualitative research. The focus of the former is on using numbers to represent concepts or opinions, while the latter is focused on making observations and using words to express perceived reality. BE embodies various domains, subjects, or topics, each of which is advanced by building a unique body of knowledge through quantitative and qualitative research. Mapping and understanding these large bodies of knowledge in an *objective* manner requires science mapping (SM) methodology.

Thus, in recent years, SM has been given wide attention in BE, where researchers have used it to analyse various bibliometric networks (e.g. keyword co-occurrence networks). The main benefit of SM of a BE research domain is in achieving comprehensive understanding of how the domain is structured and how it evolves dynamically. Specifically, SM is useful for identifying: the main areas of research in the domain; intellectual milestones in developing core specialities; evolutionary stages of the key specialities involved; and the dynamics of transitions from one speciality to another (Chen, 2017). All this knowledge could inform decisions made in research, policy, and practice.

What is science mapping?

Bibliometrics provides techniques for assessing research outputs quantitatively (Cobo *et al.*, 2011a). Bibliometrics has two main pillars: performance analysis and SM (Noyons *et al.*, 1999). While performance analysis is used to assess an activity of scientific *actors* (e.g. researchers) and the impact of this activity, based upon bibliographic data (Noyons *et al.*, 1999), SM is used to display the dynamic and structural features of scientific research (Börner *et al.*, 2003). In this chapter, SM is explored. SM is an effective methodology for finding "representations of the intellectual connections within the dynamically changing system of scientific knowledge" (Small, 1997, p. 275) by creating, analysing, and visualising bibliometric networks. These networks show how scientific specialities, fields, disciplines, and authors or articles are conceptually, socially, and intellectually structured and related to one another as illustrated by their relative locations and physical proximity (Small, 1999; Cobo *et al.*, 2011b). This illustration is similar to how geographic maps display the relations of physical or political features of the earth (Small, 1999). The focus of SM is on monitoring and delineating research areas of scientific realms to ascertain their (cognitive) structures, their evolutions as well as the core actors involved (Noyons *et al.*, 1999).

Focus of this chapter

The purpose of this chapter is to introduce the application of SM in BE research. Essentially, this introduction contributes to addressing the broader question of how to use SM in BE, by addressing the following key questions: (1) what is SM?; (2) why is it useful for BE research? (3) how can it be done effectively?; (4) which bibliographic databases can be used for SM in BE?; (5) what is the acceptable sample size?; (6) which SM analysis types and units can be used?; and (7) what are the commonly used software packages and what combinations of them can be implemented to perform high-quality, SM-based research? The chapter includes a brief tutorial to show how three software packages, VOSviewer, CiteSpace and Gephi, can be applied together to conduct robust/deep SM-based research.

Having described SM and its usefulness for BE research earlier, a survey of attempts to apply SM in the BE field, with a focus on the research domains addressed, is presented in the next section, followed by discussions about issues in using SM in this field with regards to data sources, sample size, analysis types and units, and software. The tutorial on the co-operative use of VOSviewer, CiteSpace and Gephi is then presented, followed by the conclusions from this chapter.

Historical survey

In this section, a brief overview of some of the attempts to apply SM in BE research is given. It is acknowledged generally that, at the time of writing, applying SM in BE research is a relatively recent development. One of the first attempts was by Jide *et al.* (2015), who presented an SM analysis of the literature about the occupational mobility of construction workers. Subsequently, SM has been applied in analysing construction and demolition (C&D) waste research (Liu *et al.*, 2017; Chen *et al.*, 2018; Jin *et al.*, 2019a). There have been several SM-based studies in the BE field that were focused on the research area of building information modelling (BIM) (Li *et al.*, 2017; He *et al.*, 2017; Oraee *et al.*, 2017;

Zhao, 2017; Chen and Man, 2018; Hosseini *et al.*, 2018; Chihib *et al.*, 2019; Jin *et al.*, 2019b; Saka and Chan, 2019a,b). While research on the application of BIM to BE in general was analysed in these studies, in other studies, SM was used to analyse research on the role of BIM in specific BE domains. For example, Yin *et al.* (2019) undertook an SM analysis of research on BIM for offsite construction. So far, the focus of using SM in BE has been predominantly on the literature about BIM. However, limited attention has been paid to integrating BIM with other digital technologies in addressing BE issues. One of the few attempts was by Wang *et al.* (2019), who presented a bibliometric analysis of the research on integrating BIM and GIS in sustainable BE. Other BE research domains in which applications of SM are found include green building (Darko *et al.*, 2019), embodied energy of buildings (Zeng and Chini, 2017), value management (Ekanayake *et al.*, 2019), mental health (Nwaogu *et al.*, 2019), and lean construction (He and Wang, 2015). It was found that there is scope for further use of SM in these domains.

SM data sources for BE

An SM analysis begins with retrieval of relevant bibliographic data. Today, there are many online bibliographic databases in which scientific documents together with their citations are stored. These bibliographic data sources make it possible to search and retrieve information concerning most scientific fields (Cobo *et al.*, 2011b). Nonetheless, choosing an appropriate source for data retrieval, which contains data that could offer answers to the questions to be explored, is important (Börner *et al.*, 2003). Chen (2017) and Cobo *et al.* (2011b) noted that the most commonly used bibliographic databases include Web of Science (WoS), Scopus, Google Scholar, and PubMed. The results of an analysis of 20 selected SM applications in BE reinforced this observation where WoS and Scopus have dominated current SM-based research programmes in BE, as shown in Table 9.1, with the main reasons being:

- WoS is the most authoritative database for studying literature in many fields because it contains the most prestigious and important journals of influence in the world.
- Scopus has a wider scientific publications coverage and more recent publications than other databases, such as WoS.

It is logical that PubMed is not popular within this field because it is principally a biomedical database. Although Google Scholar can be used in this field, "downloading large datasets from Google Scholar is difficult and a dump of the entire dataset is not available" (Cobo *et al.*, 2011b, p. 1383). This might be a reason for BE researchers to avoid using Google Scholar for SM. Another database that can be employed is CNKI (Chen and Man, 2018).

Funding or grants data and patent data can also be used for SM, but that is beyond the scope of this chapter; thus databases for these are not considered.

Table 9.1 Selected SM applications in BE

Study	Research domain	Data sources	Sample size[a]	Timespan	Document types	Analysis types	Analysis units	Software
Jide et al. (2015)	Construction workers' occupational mobility	WoS	190	1986–2014	Articles	Co-citation; co-occurrence; citation burst; clustering	Document co-citation (also known as cited references); author co-citation (also known as cited authors); keywords	CiteSpace
Liu et al. (2017)	C&D waste	WoS	857	2000–2016	Articles; reviews; proceedings papers; editorials	Co-authorship; co-citation; clustering	Institutions; journal co-citation (also known as cited sources); document co-citation	CiteSpace
Chen et al. (2018)	C&D waste	WoS	261	2006–2018	Articles	Co-citation; co-occurrence; citation burst; clustering	Document co-citation; keywords	CiteSpace
Jin et al. (2019a)	C&D waste	Scopus	370	2009–2018	Articles	Direct citation; co-occurrence; co-authorship	Journals; keywords; authors; documents; countries	VOSviewer
Zhao (2017)	BIM	WoS	614	2005–2016	Articles	Co-citation; direct citation; co-occurrence; co-authorship	Countries; institutions; subject categories; keywords; journal co-citation; author co-citation; document co-citation; clustering	CiteSpace
Li et al. (2017)	BIM	WoS	1,874	2004–2015	Articles; reviews; proceedings papers	Co-occurrence; co-citation; clustering; citation burst	Keywords; document co-citation; documents	CiteSpace
He et al. (2017)	BIM	WoS; Scopus	126	2007–2015	Articles	Co-occurrence; citation burst; clustering	Keywords	CiteSpace
Oraee et al. (2017)	BIM	Scopus	1,031	2006–2016	Articles; reviews	Direct citation; co-occurrence	Documents; keywords; journals	VOSviewer; Gephi
Chen and Man (2018)	BIM	WoS; CNKI	8,897	2003–2017	Articles	Co-authorship; co-occurrence	Authors; keywords	VOSviewer
Jin et al. (2019b)	BIM	Scopus	276	2008–2018	Articles	Direct citation; co-occurrence	Journals; documents; keywords	VOSviewer

Study	Topic	Database	Sample size	Period	Document type	Analysis	Units	Software
Saka and Chan (2019a)	BIM	Scopus	93	2010–2018	All	Co-authorship; co-occurrence; co-citation	Keywords; authors; author co-citation; document co-citation	VOSviewer
Saka and Chan (2019b)	BIM	WoS	914	2006–2017	Articles	Co-authorship; co-citation; co-occurrence	Keywords; authors; author co-citation; document co-citation	VOSviewer; CiteSpace
Chihib et al. (2019)	BIM	Scopus	4,307	2003–2018	All	Co-authorship; co-occurrence; clustering	Countries; keywords	VOSviewer
Wang et al. (2019)	BIM-GIS integration in sustainable BE	WoS	76	2008–2018	Articles; proceedings papers	Co-occurrence; co-authorship	Keywords; authors	VOSviewer
Wuni et al. (2019)	Green building	Scopus	1,147	1992–2018	Articles	Direct citation; co-occurrence; bibliographic coupling; co-authorship	Journals; keywords; documents; countries; authors	VOSviewer
Nwaogu et al. (2019)	Mental health	WoS; Scopus	145	1974–2018	Articles	Co-citation; co-authorship; co-occurrence; clustering; bibliographic coupling; citation burst	Author co-citation; document co-citation; keywords; authors; institutions; countries; documents	CiteSpace
He and Wang (2015)	Lean construction	WoS; Scopus	621	1995–2014	Articles	Co-authorship; co-occurrence	Countries; keywords	CiteSpace
Ekanayake et al. (2019)	Value management	WoS	1,139	1990–2017	Articles; reviews; proceedings papers	Co-citation; clustering; co-occurrence; citation burst	Document co-citation; keywords; documents	CiteSpace
Cristino et al. (2018)	Energy efficiency in buildings	Scopus	513	1980–2016	Articles	Clustering	Keywords	VOSviewer
Zeng and Chini (2017)	Embodied energy of buildings	WoS	398	1996–2015	Articles; reviews	Co-occurrence; clustering; citation burst	Keywords	CiteSpace; Gephi

Source: Original.

Note: [a] Average sample size = summation of sample sizes (23,849) divided by 20 = 1,192.45.

Sample size

An essential aspect of using SM is to be *comprehensive* in the number of documents to be analysed. It is widely known that SM is a quantitative method proposed to overcome certain limitations of manual literature analysis, one of which relates to the number of documents that can be analysed (Yalcinkaya and Singh, 2015). However, from Table 9.1, it is noted that some SM applications in BE still involved sample sizes of less than 200 documents, for example, which still could be analysed manually. This could be because of a lack of standards guiding the determination of an adequate SM sample size in the field. Based on the average sample size in Table 9.1, it is recommended that a sample of 1,000 or more documents might be considered to be acceptable or adequate for SM; while below 1,000 might require the researcher to gather more data (some ways to do this are discussed later in this section). The robustness of the literature sample is related directly to the robustness of the results.

SM can be employed to analyse "huge amounts of data" (Börner *et al.*, 2003, p. 209) that might not be possible to analyse manually because SM data analysis is computer aided. Although there is no specific definition of what "huge amounts of data" constitutes, considering the data of most SM examples in the bibliometrics field, which was where SM originated (e.g. 36,000 documents in Small, 1999; and 25,242 in van Eck and Waltman, 2014a), it is sound to assume that "huge amounts of data" refers to datasets consisting of thousands of documents. It would be useful for BE researchers to follow this practice in conducting SM-based research. However, if datasets are limited to hundreds of documents, it is advisable that such datasets should consist of 500–999 documents. Anything below 500 documents could be deemed to be a weak or unacceptable sample size for SM.

While acknowledging that the sample size will depend on the amount of publications available in the particular research domain, this should not become a reason to compromise or undermine the comprehensiveness and robustness SM can offer in secondary research. There are several ways in which to improve the sample size as follows:

- Use a comprehensive list of keywords for the literature search. An example of this is demonstrated later in the "Co-operative use of VOSviewer, CiteSpace, and Gephi: a brief tutorial" section.
- Employ the citation expansion method to expand the original dataset by merging it with a dataset of documents within its citation record (Li *et al.*, 2017).
- Use a comprehensive timespan (e.g. all years to the present) for the literature search.
- Include multiple document types (e.g. journal and conference articles).
- Employ multiple bibliographic databases (e.g. WoS and Scopus).

In any case, data pre-processing is necessary to fix errors, such as irrelevant documents, and thus improve the quality of data, which will affect the quality of the SM results. Cobo *et al.* (2011b) discuss SM data pre-processing in detail.

If all possible ways have been attempted and yet an adequate sample size of 1,000 or more documents or 500–999 documents cannot be reached, then it is necessary to consider asking: why conduct SM when the body of literature in the particular research domain is still young, given that having a huge dataset is critical for robust SM?

SM analysis types and units

There are several types of SM analysis for identifying the dynamic evolution and knowledge structure of a domain through constructing, analysing, and visualising bibliometric networks. In VOSviewer software, for instance, these types include co-authorship, citation, co-occurrence, co-citation, bibliographic coupling, and clustering analyses which are presented in Table 9.2, together with their applications. Extra types such as citation burst analysis can be found in software such as CiteSpace.

In BE, the most commonly used or most important analysis types include co-occurrence, co-authorship, clustering, co-citation, citation burst, and citation analyses; with the most commonly used analysis units including keywords, journals, authors, documents, institutions, document co-citation or cited references, and author co-citation or cited authors (Table 9.1). While bibliographic coupling can also be used, it has yet to attract the level of attention that other types of analysis have attracted. Additionally, journal co-citation or cited sources as an analysis unit has yet to receive much attention. For the sake of brevity, van Eck and Waltman (2014a) and Börner *et al.* (2003) are cited for further discussions of SM analysis types and units.

Table 9.2 VOSviewer-based SM analysis types and their application

Analysis types	Application
Co-authorship	Identifying influential researchers, institutions, or countries based on collaborations
Co-occurrence	Identifying major research interests, areas, or topics based on keywords
Citation	Identifying influential journals, researchers, institutions, countries, or articles based on direct citations
Bibliographic coupling	Identifying influential journals, researchers, institutions, countries, or articles based on shared references
Co-citation	Identifying influential journals, researchers, or articles based on co-citations
Clustering	Identifying groups of related topics, researchers, institutions, countries, journals, or articles

Source: Original.

SM software packages

Various software packages have been developed to conduct SM analysis, including:

- CiteSpace (http://cluster.cis.drexel.edu/~cchen/citespace/; Chen, 2014).
- VOSviewer (https://www.vosviewer.com; van Eck and Waltman, 2019).
- Sci2 (https://sci2.cns.iu.edu/user/index.php; Sci2 Team, 2009).
- BibExcel (https://homepage.univie.ac.at/juan.gorraiz/bibexcel/; Persson *et al.*, 2009).
- Ucinet (https://sites.google.com/site/ucinetsoftware/home, Borgatti *et al.*, 2002).
- SciMAT (https://sci2s.ugr.es/scimat/; Cobo *et al.*, 2012).

The aim of this chapter is not to give an overview of available SM software tools, but rather to introduce and direct the reader (especially those who are new to SM) to some of the tools. Cobo *et al.* (2011b) and van Eck and Waltman (2014a) have already provided informative overviews of software tools for SM, which include other tools not mentioned in this chapter.

Some tools (e.g. BibExcel and Ucinet) were developed specifically for bibliographic data processing, after which the results must be imported into visualisation software such as Gephi (https://gephi.org; Bastian *et al.*, 2009) and Pajek (http://pajek.imfm.si; Batagelj and Mrvar, 1998) before the network can be visualised and analysed. Note that because Gephi and Pajek are specifically developed for network visualisation and analysis, they have no functionality for data processing. However, tools such as CiteSpace, VOSviewer, Sci2, and SciMAT can perform both data processing and network visualisation and analysis. Tools possess different capabilities and strengths and present complementary features. Therefore, Cobo *et al.* (2011b, p. 1383) recommend "to take their synergies to perform a complete SM analysis". This has not been sufficiently addressed in BE. While CiteSpace, VOSviewer, and Gephi have been the most popular or most important tools in the BE domain, only limited attempts have been made towards using multiple tools, e.g. VOSviewer+CiteSpace (Saka and Chan, 2019b), CiteSpace+Gephi (Zeng and Chini, 2017), and VOSviewer+Gephi (Oraee *et al.*, 2017) in one study (Table 9.1).

Much of the popularity gained by CiteSpace, VOSviewer, and Gephi within this field could be accredited to some of their individual strengths. CiteSpace can be used to build and visualise bibliometric networks dynamically to show how the studied domain has evolved over time and provide users with a wider range of visualisation and analysis options than other software, such as VOSviewer. A key strength of VOSviewer lies in its being easy to use with special attention paid to the graphical presentation of the bibliometric networks; it also has capacity for large networks. Unlike other software (e.g. CitNetExplorer; https://www.citnetexplorer.nl; van Eck and Waltman, 2014b) that can be used to visualise and analyse only one bibliometric network type, i.e. citation networks of publications, Gephi can be used to visualise and analyse all network types.

In the next section, a tutorial is presented that shows how VOSviewer, CiteSpace, and Gephi can be combined for robust and deep SM analysis. Although other possible synergies among the different software exist, the combination of these three software packages is one of the highest synergies achieved in the BE field so far. In fact, a complete, thorough, and deep SM analysis of any domain requires combined use of different software (Cobo *et al.*, 2011b). Such an approach affords the extraction of all the useful knowledge and diverse perspectives about the domain embedded in the dataset.

Combined use of VOSviewer, CiteSpace, and Gephi: a brief tutorial

The aim of this tutorial is to facilitate and promote the idea of using different software packages in combination to perform complete, thorough, and deep SM in BE. To this end, an example of how VOSviewer, CiteSpace, and Gephi can be used in combination by following the steps below (Darko *et al.*, 2020) is demonstrated.

1 Determine the research domain you wish to study. For example, in this tutorial, the AI in the AEC industry research domain is selected.
2 Establish a list of keywords for the literature search. This step is important because the comprehensiveness and robustness of the search keywords list affects these aspects of the whole SM activity from sample size to final results. As an example, results (sample sizes) obtained from searches in Scopus with three different possible keywords lists for the research domain considered in this tutorial are shown in Table 9.3. All the searches were run at the same time on 26 January 2020. Figures 9.1 and 9.2 show examples of how the final results/networks might be affected in terms of comprehensiveness and robustness. Note that the two figures were created using the same process. For instance, they were both created using VOSviewer, with "author keywords" as the analysis unit, fractional counting method was used, and the minimum number of occurrences a keyword should have in order to be included in the network was set to five for both figures. The only difference was the dataset.
3 Choose the bibliographic database(s) to be used and download the bibliographic data from there using the keywords list established in Step 2. For example, in this tutorial, Scopus was used and the data were downloaded in comma-separated values (CSV) format. In Scopus, bibliographic data for only 2,000 or less documents can be downloaded at a time. If more than 2,000 documents are available, downloading can be done in batches based upon journals, for example.
4 Once the dataset is ready, download and launch VOSviewer, CiteSpace, and Gephi. They are all freely downloadable (which is a reason for their wide use) from https://www.vosviewer.com; http://cluster.cis.drexel.edu/~cchen/citespace/; and https://gephi.org, respectively.

Table 9.3 Keywords and literature search results

Query string	Number of keywords	Search results (sample size)
"Artificial intelligence" AND "Architecture, Engineering and Construction industry" AND [LIMIT-TO (SUBJAREA, "ENGI")] AND [LIMIT-TO (DOCTYPE, "ar")]	2	58 documents
"Artificial intelligence" OR "Machine intelligence" OR "Machine learning" OR "Expert systems" OR "Genetic algorithms" OR "Artificial neural networks" OR "Artificial general intelligence" OR "Case-based reasoning" AND "Architecture, Engineering and Construction industry" AND [LIMIT-TO (SUBJAREA, "ENGI")] AND [LIMIT-TO (DOCTYPE, "ar")]	9	102 documents
"Artificial intelligence" OR "Machine intelligence" OR "Machine learning" OR "Expert systems" OR "Genetic algorithms" OR "Neural networks" OR "Case-based reasoning" OR "Data mining" OR "fuzzy logic" OR "Fuzzy sets" OR "Expert systems" OR "Robotics" OR "Knowledge-based systems" OR "Support vector machines" OR "Deep learning" OR "Artificial general intelligence" OR "Computational intelligence" AND "Construction industry" OR "Civil engineering" OR "Structural engineering" OR "Architectural engineering" OR "Construction engineering" OR "Construction management" OR "Construction engineering and management" AND [LIMIT-TO (SUBJAREA, "ENGI")) AND (LIMIT-TO (DOCTYPE, "ar")]	24	53,924 documents

Source: Original.

5   Use VOSviewer to build a keyword co-occurrence network (Figures 9.1 and 9.2 are examples). However, a drawback of VOSviewer is that it does not have data pre-processing modules, so there might be duplicate items (keywords in this case). Although, at a point, VOSviewer offers the option to remove items manually from the network, the option was not selected for Figures 9.1 and 9.2 to emphasise the importance of data pre-processing. If the option to remove items manually is selected, this can be a tedious task, especially with numerous keywords, and if any item is missed before moving to the next step, it is necessary to restart the whole analysis process. There are duplicate items, such as BIM, building information modelling, and building information modelling, in Figure 9.1. Duplicate items can be merged using a thesaurus file, but this is still a manual, time-consuming process. Before the analysis in VOSviewer, de-duplication could be done using CoPalRed software (Cobo *et al.*, 2011b).

6   Use CiteSpace to perform cluster analysis with document co-citation as the analysis unit (other analysis units can also be used). In the space of this

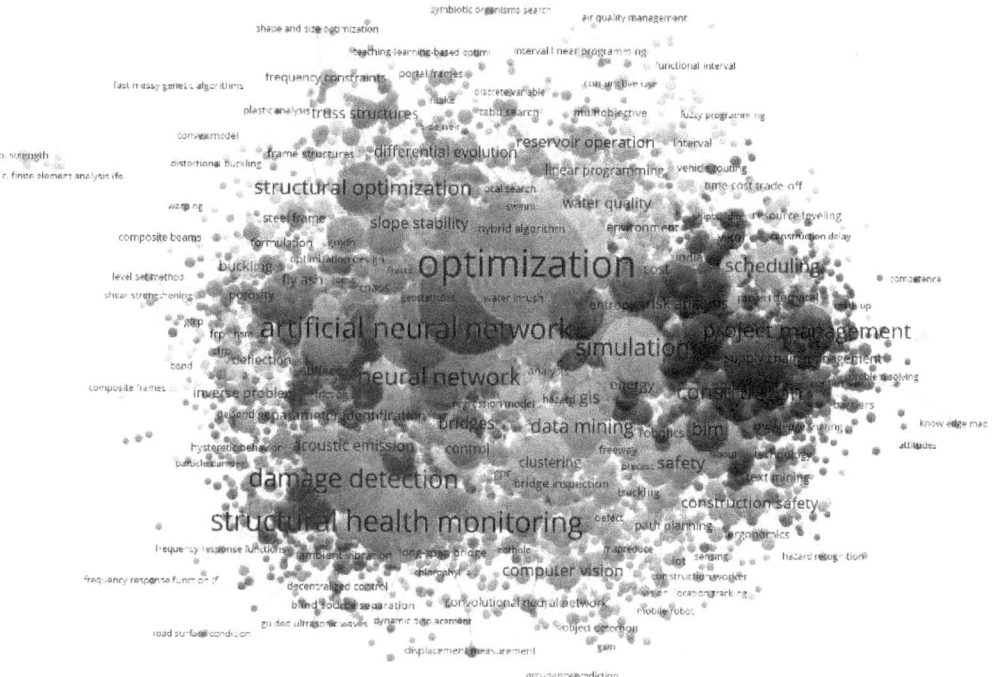

Figure 9.1 VOSviewer visualisation of a 4,297 author keywords co-occurrence network based on the dataset of 53,924 documents.

Figure 9.2 VOSviewer visualisation of a five author keywords co-occurrence network based on the dataset of 102 documents.

chapter, it is not possible to show more figures and tables; more can be found in Darko *et al.* (2020). While VOSviewer can be used also for cluster analysis (for instance, the different colours in Figure 9.1 represent different clusters of the keywords), "CiteSpace provides more precise ways to identify groupings, or clusters, using the clustering function" (Chen, 2014, p. 14).

7   Use CiteSpace again to perform citation burst analysis to detect how the most active areas or features of the research domain have evolved over time. The key strength of CiteSpace is that it has the capacity to detect time-based evolutions of features of the domain by automatically slicing the dataset into different time periods for the analysis. Citation burst analysis can be performed over many analysis units, including keywords, documents, authors, and institutions. In each case, active areas and/or emerging trends can be revealed. An example of citation burst analysis over keywords is provided in Darko *et al.* (2020).

8   Close CiteSpace and go back to VOSviewer. Use VOSviewer to conduct a direct citation analysis of sources (or journals). Save the generated network in graph modelling language (GML) format and then submit this to Gephi for further analysis. The benefit of this combined approach is that it leads to higher quality visualisation of the network, as Gephi affords further options and capabilities. For example, it offers a choice between directed, undirected, and mixed network types. Moreover, sometimes VOSviewer does not show the names of some nodes in the network. This is evident in Figure 9.1, for example. The example is based on a large network, but sometimes, even with relatively small networks, this problem still occurs and can be fixed with Gephi. VOSviewer also does not always present the full names of nodes (Jin *et al.*, 2019b), which is another problem that can be solved with Gephi. Furthermore, VOSviewer can only show lower-case letters, so "IT", for example, might appear as "it" (Jin *et al.*, 2019c). Gephi can be used to solve this issue because it makes it possible to edit the names of nodes. Gephi also allows further manipulation of the network to achieve the desired shape and appearance. Direct citation analysis can be performed for documents, authors, institutions, or countries as well.

9   Use VOSviewer to perform co-authorship analysis of institutions and of countries, which makes it possible to explore collaborations among research institutions and among countries. Save the generated networks in GML format and then submit them to Gephi for further analysis. Co-authorship analysis can be done for authors too.

10 After completing the above steps, note that the software assist the user to produce results and networks based on the research dataset, but cannot interpret and draw conclusions from the results. Thus, the final step is to interpret and draw conclusions from the results. The quality and depth of the interpretation and conclusions depend on the user's knowledge and experience. The key point is to extract useful knowledge that could well inform further research, policy, and practice decisions.

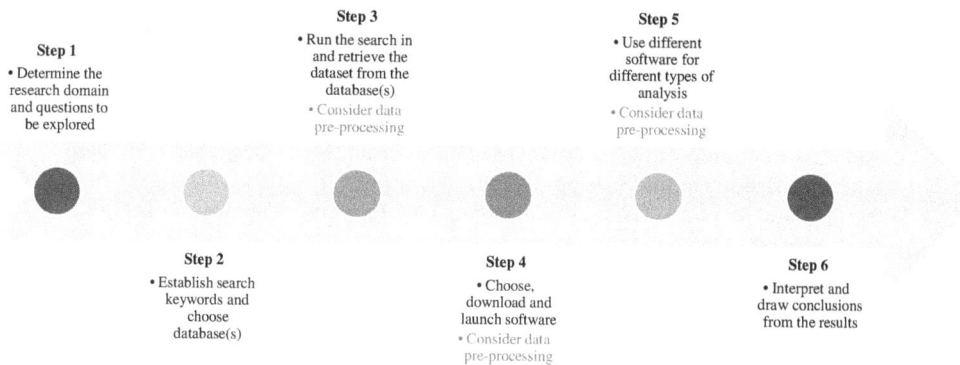

Figure 9.3 Generic six-step procedure for SM-based research.

For SM-based studies in BE, where this combined use of VOSviewer, CiteSpace, and Gephi was employed, refer to Darko *et al.* (2019, 2020) and Hosseini *et al.* (2018). This tutorial gives an example of generic steps to follow to conduct SM through the combined use of these three software packages, but does not show how to use the software. This can be found in manuals for VOSviewer (van Eck and Waltman, 2019), CiteSpace (Chen, 2014), and Gephi (Bastian *et al.*, 2009). While the tutorial was focused only on some synergies between VOSviewer, CiteSpace, and Gephi, other synergies among these and/or other software could also be explored in BE (Cobo *et al.*, 2011a,b). Hence, we propose a generic six-step procedure (Figure 9.3) that could be adopted/adapted for any SM-based research (especially those conducted through combined use of different software).

Conclusions

BE domains, subjects, or topics are advanced by building unique bodies of knowledge through quantitative and qualitative research. Mapping and understanding these large bodies of knowledge in an *objective* manner requires SM. In this chapter, the topic of applying SM in BE research is introduced. This introduction contributes to addressing the question of how to use SM in BE. Two main recommendations are to: (1) use *large* sample sizes (1,000 or more documents) for SM in BE because SM quality and robustness depends on sample size robustness; and (2) widely adopt the combined use of various software to conduct deep, thorough, and complete SM activities in BE research. This approach not only assists to extract all the useful knowledge from the dataset, but also to compensate for each software package's weaknesses by combining their strengths. It is hoped that this chapter will help to promote widespread use of this integrated approach and help researchers and other interested stakeholders to conduct quality research using SM.

References

Amaratunga, D., Baldry, D., Sarshar, M. and Newton, R. (2002). Quantitative and qualitative research in the built environment: Application of "mixed" research approach. *Work Study*, 51(1), pp. 17–31.

Bastian, M., Heymann, S. and Jacomy, M. (2009). Gephi: An open source software for exploring and manipulating networks. *Third International AAAI Conference on Weblogs and Social Media*, San Jose, CA, pp. 361–362.

Batagelj, V. and Mrvar, A. (1998). Pajek-program for large network analysis. *Connections*, 21(2), pp. 47–57.

Borgatti, S.P., Everett, M.G. and Freeman, L.C. (2002). *Ucinet for Windows: Software for Social Network Analysis*, Analytic Technologies, Harvard, MA.

Börner, K., Chen, C. and Boyack, K.W. (2003). Visualizing knowledge domains. *Annual Review of Information Science and Technology*, 37(1), pp. 179–255.

Chen, C. (2014). CiteSpace Manual. Available at: http://cluster.ischool.drexel.edu/~cchen/citespace/CiteSpaceManual.pdf.

Chen, C. (2017). Science mapping: A systematic review of the literature. *Journal of Data and Information Science*, 2(2), pp. 1–40.

Chen, J. and Man, Q. (2018). A critical review on BIM research process and future trends at home and abroad. *ICCREM 2018: Innovative Technology and Intelligent Construction*, Charleston, SC, pp. 104–112.

Chen, J., Su, Y., Si, H. and Chen, J. (2018). Managerial areas of construction and demolition waste: A scientometric review. *International Journal of Environmental Research and Public Health*, 15(11), p. 2350.

Chihib, M., Salmerón-Manzano, E., Novas, N. and Manzano-Agugliaro, F. (2019). Bibliometric maps of BIM and BIM in universities: A comparative analysis. *Sustainability*, 11(16), p. 4398.

Cobo, M.J., López-Herrera, A.G., Herrera-Viedma, E. and Herrera, F. (2011a). An approach for detecting, quantifying, and visualizing the evolution of a research field: A practical application to the fuzzy sets theory field. *Journal of Informetrics*, 5(1), pp. 146–166.

Cobo, M.J., López-Herrera, A.G., Herrera-Viedma, E. and Herrera, F. (2011b). Science mapping software tools: Review, analysis, and cooperative study among tools. *Journal of the American Society for Information Science and Technology*, 62(7), pp. 1382–1402.

Cobo, M.J., López-Herrera, A.G., Herrera-Viedma, E. and Herrera, F. (2012). SciMAT: A new science mapping analysis software tool. *Journal of the American Society for Information Science and Technology*, 63(8), pp. 1609–1630.

Cristino, T.M., Neto, A.F. and Costa, A.F.B. (2018). Energy efficiency in buildings: Analysis of scientific literature and identification of data analysis techniques from a bibliometric study. *Scientometrics*, 114(3), pp. 1275–1326.

Darko, A., Chan, A.P.C., Adabre, M.A., Edwards, D.J., Hosseini, M.R. and Ameyaw, E.E. (2020). Artificial intelligence in the AEC industry: Scientometric analysis and visualization of research activities. *Automation in Construction*, 112, p. 103081.

Darko, A., Chan, A.P.C., Huo, X., and Owusu-Manu, D.G. (2019). A scientometric analysis and visualization of global green building research. *Building and Environment*, 149, pp. 501–511.

Ekanayake, E.M.A.C., Shen, G. and Kumaraswamy, M.M. (2019). Mapping the knowledge domains of value management: A bibliometric approach. *Engineering, Construction and Architectural Management*, 26(3), pp. 99–514.

He, Q.H. and Wang, G. (2015). Hotspots evolution and frontier analysis of lean construction research—Integrated scientometric analysis using the Web of Science and Scopus databases. *Frontiers of Engineering Management*, 2(2), pp. 141–147.

He, Q., Wang, G., Luo, L., Shi, Q., Xie, J. and Meng, X. (2017). Mapping the managerial areas of building information modeling (BIM) using scientometric analysis. *International Journal of Project Management*, 35(4), pp. 670–685.

Hosseini, M.R., Maghrebi, M., Akbarnezhad, A., Martek, I. and Arashpour, M. (2018). Analysis of citation networks in building information modeling research. *Journal of Construction Engineering and Management*, 144(8), p. 04018064.

Jide, S., Qi, N. and Liangfa, S. (2015). Analysis on current situation and development trend of construction worker's occupational mobility. *Open Construction and Building Technology Journal*, 9, pp. 303–310.

Jin, R., Yuan, H. and Chen, Q. (2019a). Science mapping approach to assisting the review of construction and demolition waste management research published between 2009 and 2018. *Resources, Conservation and Recycling*, 140, pp. 175–188.

Jin, R., Zou, Y., Gidado, K., Ashton, P. and Painting, N. (2019b). Scientometric analysis of BIM-based research in construction engineering and management. *Engineering, Construction and Architectural Management*, 26(8), pp. 1750–1776.

Jin, R., Zou, P.X., Piroozfar, P., Wood, H., Yang, Y., Yan, L. and Han, Y. (2019c). A science mapping approach based review of construction safety research. *Safety Science*, 113, 285–297.

Li, X., Wu, P., Shen, G.Q., Wang, X. and Teng, Y. (2017). Mapping the knowledge domains of Building Information Modeling (BIM): A bibliometric approach. *Automation in Construction*, 84, pp. 195–206.

Liu, Y., Sun, T. and Yang, L. (2017). Evaluating the performance and intellectual structure of construction and demolition waste research during 2000–2016. *Environmental Science and Pollution Research*, 24(23), pp. 19259–19266.

Noyons, E., Moed, H. and Van Raan, A. (1999). Integrating research performance analysis and science mapping. *Scientometrics*, 46(3), pp. 591–604.

Nwaogu, J.M., Chan, A.P.C., Hon, C.K. and Darko, A. (2019). Review of global mental health research in the construction industry. *Engineering, Construction and Architectural Management*. doi:10.1108/ECAM-02-2019-0114.

Oraee, M., Hosseini, M.R., Papadonikolaki, E., Palliyaguru, R. and Arashpour, M. (2017). Collaboration in BIM-based construction networks: A bibliometric-qualitative literature review. *International Journal of Project Management*, 35(7), pp. 1288–1301.

Persson, O., Danell, R. and Schneider, J.W. (2009). Åström, F., Danell, R., Larsen, B., and Schneider, J. (Eds.), How to use Bibexcel for various types of bibliometric analysis. *Celebrating scholarly Communication Studies: A Festschrift for Olle Persson at his 60th Birthday*, International Society for Scientometrics and Informetrics (ISSI), Leuven, pp. 5, 9–24.

Saka, A.B. and Chan, D.W.M. (2019a). A scientometric review and metasynthesis of building information modelling (BIM) research in Africa. *Buildings*, 9(4), p. 85.

Saka, A.B. and Chan, D.W.M. (2019b). A global taxonomic review and analysis of the development of BIM research between 2006 and 2017. *Construction Innovation*, 19(3), pp. 465–490.

Sci2 Team. (2009). Science of Science (Sci2) Tool. Available at: https://sci2.cns.iu.edu/user/index.php.

Small, H. (1997). Update on science mapping: Creating large document spaces. *Scientometrics*, 38(2), pp. 275–293.

Small, H. (1999). Visualizing science by citation mapping. *Journal of the American society for Information Science*, 50(9), pp. 799–813.

van Eck, N.J. and Waltman, L. (2014a). Visualizing bibliometric networks. In: Y. Ding, R. Rousseau and D. Wolfram (eds). *Measuring Scholarly Impact*, Springer, Cham, pp. 285–320.

van Eck, N.J. and Waltman, L. (2014b). CitNetExplorer: A new software tool for analyzing and visualizing citation networks. *Journal of Informetrics*, 8(4), pp. 802–823.

van Eck, N.J. and Waltman, L. (2019). VOSviewer Manual. Available at: https://www.vosviewer.com/documentation/Manual_VOSviewer_1.6.11.pdf.

Wang, H., Pan, Y. and Luo, X. (2019). Integration of BIM and GIS in sustainable built environment: A review and bibliometric analysis. *Automation in Construction*, 103, pp. 41–52.

Wuni, I.Y., Shen, G.Q. and Osei-Kyei, R. (2019). Scientometric review of global research trends on green buildings in construction journals from 1992 to 2018. *Energy and Buildings*, 190, pp. 69–85.

Yalcinkaya, M. and Singh, V. (2015). Patterns and trends in building information modeling (BIM) research: A latent semantic analysis. *Automation in Construction*, 59, pp. 68–80.

Yin, X., Liu, H., Chen, Y. and Al-Hussein, M. (2019). Building information modelling for off-site construction: Review and future directions. *Automation in Construction*, 101, pp. 72–91.

Zeng, R. and Chini, A. (2017). A review of research on embodied energy of buildings using bibliometric analysis. *Energy and Buildings*, 155, pp. 172–184.

Zhao, X. (2017). A scientometric review of global BIM research: Analysis and visualization. *Automation in Construction*, 80, pp. 37–47.

10 Bibliometric analysis for reviewing published studies in the built environment

Liyuan Wang, Ruoyu Jin, and Joseph Kangwa

Introduction

Literature review is considered to be crucial in synthesising past research findings as well as in introducing new concepts, theories, or paradigms. In any of these activities, the aim is to sustain a degree of professional judgement and expertise (Rousseau, 2012). Traditionally, the manual-based literature review has been widely adopted, primarily to unveil the trend of a given research domain. The manual review is usually based on a limited sample of literature. For example, the sample could be from influential or mainstream academic journals which assist researchers in supporting their own points of view further. A potential limitation of the manual review is its subjectivity, as mentioned by Song *et al.* (2016) and Hosseini *et al.* (2018). The number of academic publications is increasing, which is a phenomenon that is critical in determining any rapid paradigm shift within well-established models and bodies of theories. It is less feasible today to remain current given the breadth and depth of researched knowledge and the ever-expanding channels for disseminating new findings through open access (Aria and Cuccurullo, 2017). Therefore, bibliometric analysis is being adopted widely to assist in the literature review process by offering a less subjective, more reliable, and time-saving approach to search mainstream keywords and other citation-related metrics within a defined research domain. For example, in a review-based study, bibliometric analysis can assist in answering three general research questions, namely: (1) within a defined domain, what are the current mainstream topics?; (2) what are the notable gaps or limitations in existing studies?; and (3) what are the recommended or plausible areas dictating future research directions? Examples of adopting bibliometric analysis, which address the three research questions can be found in some existing literature within the built environment research domain, such as off-site construction (OSC; Jin *et al.*, 2018), infrastructure management (Jiang *et al.*, 2019), and sustainable transport (Zhao *et al.*, 2020).

This chapter aims to introduce the methodology of conducting a review-based study using secondary data. The secondary data are derived from academic publications, including journal articles and conference proceedings. The concept of bibliometrics first is defined as well as widely used databases of literature (e.g. Scopus), followed by the commonly adopted bibliometric analysis tools such as

VOSviewer (van Eck and Waltman, 2010). In describing the general workflow of the science mapping approach, an in-depth discussion beyond the bibliometric analysis is recommended also in the chapter by incorporating the researchers' own knowledge or expertise in the selected research domain. It is suggested that as an aid to synthesising information, the bibliometric analysis enables researchers to pre-define research objectives, but should not be treated as the research aim. Principally, the bibliometric analysis is a tool to assist researchers to derive the answers to the research questions searching mainstream keywords or topics within the pre-defined domain. Researchers are encouraged to progress beyond the network analysis of the citation data (e.g. keywords, documents) by demystifying the selected research domain. Moving from general to specific, this chapter includes also two examples of how to conduct the science mapping approach to demonstrate the workflow of bibliometric analysis. This chapter includes the rationale of conducting bibliometric analysis and the fundamental steps to follow, including examples of how the bibliometric analysis might well assist in providing clarity to the definitions of the research objectives, particularly in literature review-based studies. The two case studies on which this chapter is based can be adopted also for other research domains in the built environment and beyond.

Definition of bibliometric analysis

In many cases, bibliometrics may be used interchangeably with other terminologies such as scientometrics and informetrics, which share the same theories, methods, technologies, and applications, but differ in relation to subject specialisms (Yang and Yuan, 2017). Specifically, bibliometrics refers to an effectual library and documentation science; invariably, scientometrics is focused on the science of science and informetrics is focused on information science (Brookes, 1990; Wang, 1998; Qiu *et al.*, 2017). The three terminologies differ in their degrees of use and recognition, but offer a convergent point for the general principles enshrined in the citation of secondary sources (Yang and Yuan, 2017). More detailed descriptions of the three terms can be found in Hood and Wilson (2001). According to Hood and Wilson (2001), the three terms overlap considerably, with bibliometrics having a longer history and being used constantly. "Bibliometrics" is the most frequently used terminology, commanding the highest rate of increase of the three. However, in order to avoid confusion, it was recommended by Yang and Yuan (2017) that the term "bibliometrics" should be adopted as a general term for scientometrics and informetrics. For this reason, in the following sections of this chapter, "bibliometric" is adopted as a generalised term to represent the feature of the bibliometric method akin to text-mining-based algorithms and the derivatives of search patterns from various data sources. It was posited further by Aria and Cuccurullo (2017) that the use of bibliometrics is gradually extending to all disciplines. For example, bibliometric analysis is now found in a variety of subjects, such as pharmacy (Burghardt *et al.*, 2020), engineering (Xu *et al.*, 2018; Vilutiene *et al.*, 2019), and management theories (He *et al.*, 2017). In the field of built environment, multiple bibliometric-analysis-based studies have found a place in academic publications, for example, OSC and volumetric-modular assembly (Hosseini *et al.*, 2018), digital construction and building information modelling (Zhao, 2017), and construction waste management (Jin *et al.*, 2019a). In these extant literature sources, a review-based approach is adapted to bibliometric analysis, which enumerates the frequency of citation-based evaluations as well as those which corroborate the most cited author, linked institution, allied journal as well as the frequency of author-generated keywords and host country. These metrics are extrapolated from a sample of constitutive literature using constructs integral to pre-defined research domains, such as OSC.

Secondary data source for conducting bibliometric analysis

In conducting the bibliometric analysis of academic literature, Scopus (Elsevier) and Web of Science (WOS; Claravite Analytics) are the two databases commonly suggested to search the secondary data. These databases are suggested because they carry the bibliometric analysis functions which project traffic flow according to the number of "reads" and "downloads" as well as consistent past reports on author-publication rates within a given discipline such as built environment (Burghardt *et al.*, 2020). It is believed that Scopus covers more journals and more recent publications than other digital sources such as WOS (Aghaei Chadegani *et al.*, 2013). Nevertheless, other databases are gaining equally in popularity such as Google Scholar. It is thus not uncommon to utilise all these search engines to have a more comprehensive coverage of literature.

Within any of these databases, different types of literature sources can be found, including journal articles, conference proceedings, and book chapters. Researchers can opt to include or exclude the types of literature sources relative to the research endeavour. Butler and Visser (2006) postulated that conference papers, although generally available in larger quantities, do not often provide as good a medium for comparative bibliometric analyses and have little to offer in this aspect compared with journal articles. Researchers in bibliometric analyses, such as Xu *et al.* (2018) and Chen *et al.* (2019), are good reference points whose findings exclude conference proceedings. However, conference proceedings can be considered, depending on the context of the review. For example, there may be more educational studies in built environment published in conference proceedings. In the review of BIM education in built environment higher education, Wang *et al.* (2020) included both journal articles and conference proceedings.

Deciding to conduct bibliometric analysis

Once the concept and data sources relating to literature review are known, it is also vital to decide when and when not to adopt bibliometric analysis. Generally, bibliometric analysis, by name and features, is applicable to a relatively larger sample of literature. The definition of "relatively larger", although not related to a standard, quantified sample size of literature, is that the sample size should be large enough. This should ensure that any derived pattern analysis or quantitative summary of frequency or citation-related metrics, such as what is the most frequently studied author keyword, are outcomes of large sample frames. For a smaller sample of literature, other review methods other than bibliometric analysis might be more appropriate, including manual review involving purely qualitative discussions. The valid sample size (for example, more than 50 documents found within the defined scope) would determine whether it is wise to conduct the bibliometric analysis. The sample size of literature would be determined by the defined scope of the study. Generally, a wider scope would result in a larger sample of literature. Below is an example of how the same topic in a given discipline could be defined from a wider scope to a more narrowed-down scope:

- Review of safety management;
- Review of construction safety management;
- Review of construction safety climate;
- Review of organisational safety culture in construction;
- Review of virtual reality for construction safety.

It can be seen from the example above that a domain (e.g. safety in the built environment) can be defined within a broader or a narrower scope. There is no recommendation or preference for selecting a broader or a narrower scope other than depending on the researchers' aim and objectives of the study. It is not uncommon to adopt multiple keywords to define the scope, for example, "virtual reality applied in construction safety management".

While it is critical to define the review scope within the domain, it is possible to have the same study start from a broader scope to access a larger size of literature sample, and then emphasise a specific topic generated from the originally broader scope. The rationale of starting from a larger sample of literature is that a bigger picture is obtained before moving to the focused scope. Furthermore, sometimes a certain pattern or highlight from the larger literature sample could be identified as new findings of the research. For example, Jin et al. (2019b) started from the two main keywords relating to BIM and building performance analysis. After the bibliometric analysis of the initial literature sample, the keyword of inter-operability was identified as the focal point. This also creates the opportunity to link bibliometric analysis with other review methods such as content analysis. Thus, adopting bibliometric analysis does not exclude other review methods, making it possible to undertake a comprehensive review of the pre-defined study scope, as in the example of Jin et al. (2019b).

Software tools for conducting bibliometric analysis

There are several, widely used software tools that have been developed to assist bibliometric analysis, including, but not limited to, VOSviewer (van Eck and Waltman, 2010), CiteSpace (Chen, 2016), and Gephi (Bastian et al., 2009). The weblinks of several software tools for bibliometric analysis are listed in Table 10.1.

Table 10.1 Summary of several software tools for visualising bibliometric networks

Software tool	URL	Main description
CitNetExplorer	https://www.citnetexplorer.nl/	A software tool for visualising and analysing citation networks of scientific publications, allowing citation networks to be imported directly from the Web of Science database
CiteSpace	http://cluster.cis.drexel.edu/~cchen/citespace/	A freely available Java application for visualising and analysing trends and patterns in scientific literature
Gephi	https://gephi.org/	A visualisation and exploration software package for all kinds of graphs and networks; open-source and free
Pajek	http://mrvar.fdv.uni-lj.si/pajek/	Analysis and visualisation of very large networks
Sci^2	https://sci2.cns.iu.edu/user/index.php	A modular toolset designed specifically for the study of science; supporting the temporal, geospatial, topical, and network analysis and visualisation of scholarly datasets at the micro (individual), meso (local), and macro (global) levels
VOSviewer	https://www.vosviewer.com/	A software tool for constructing and visualising bibliometric networks, including journals, researchers, or individual publications, which can be constructed based on citation, bibliographic coupling, co-citation, or co-authorship relations; offering text mining functionality that can be used to construct and visualise co-occurrence networks of important terms extracted from a body of scientific literature

Source: Adapted from van Eck and Waltman (2014).

As shown in Table 10.1, these software tools are typically freely downloadable for use in visualising and analysing the network (e.g. based on citation) from scientific publications, despite the variability in the strength and suitability of each tool. The aim of this chapter was not to recommend any specific tool in preference to another, but to describe the main features of a few, widely used tools (e.g. VOSviewer and CiteSpace). It is the researchers' decision to adopt any tool for the bibliometric analysis, although some generally comparative descriptions can be found from user experience. For example, compared with Pajek, Gephi is focused less on network analysis and more on network visualisation with extensive visualisation capabilities (van Eck and Waltman, 2014). CiteSpace provides dynamic visualisations to show how bibliometric networks evolve over time, and offers both graph-based and timeline-based visualisations (van Eck and Waltman, 2014). CiteSpace also allows co-citation cluster analysis to generate research themes. This timeline-based visualisation feature has been adopted widely in the review of academic publications in the built environment, for example, in the domain of OSC (Hosseini *et al.*, 2018) and in the topic of sustainable transport (Zhao *et al.*, 2020). According to van Eck and Waltman (2014), VOSviewer offers an easy-to-use option, highlighting distance-based visualisation with less functionality for analysing networks. It is also practical to adopt more than one tool to conduct the bibliometric analysis of networks from a selected sample of literature, including journal or document sources, authors or scholars, institutions, keywords, and countries or regions.

Workflow of conducting bibliometric analysis

Science mapping is the workflow involving bibliometric analysis. Aria and Cuccurullo (2017) proposed the three-step science mapping method, namely data collection, data analysis, and data visualisation. The workflow of the science mapping approach can be generalised further from the review of other literature review-based studies (Jin *et al.*, 2018; Xu *et al.*, 2018; Zhao *et al.*, 2020). Figure 10.1 illustrates the typical workflow.

According to the workflow described in Figure 10.1, bibliometrics is part of science mapping, defined by literature search and screening, bibliometric study, and further discussion. The first step in searching and screening literature from the database (e.g. WOS, or Scopus) is to use pre-defined keywords. The screening process might involve one or more steps to finalise the literature sample that falls within the defined scope of study. An illustration is included in the next section, showing a specific example. After the literature sample has been finalised, the bibliometric analysis can be applied by adopting one of the software

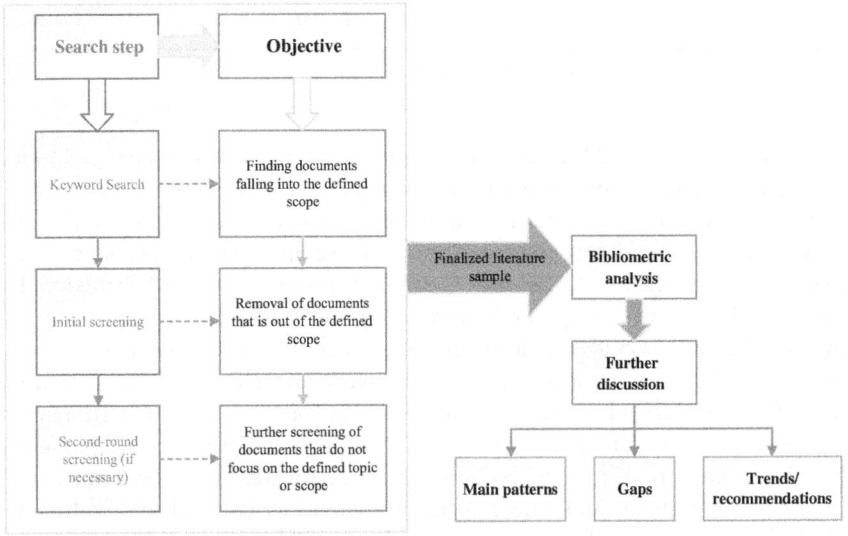

Figure 10.1 Typical workflow of conducting the bibliometric analysis in the science mapping approach (adapted from Jin *et al.*, 2019a).

tools listed in Table 10.1. A further in-depth discussion is recommended beyond the bibliometric analysis, to address the three general research questions, corresponding to the mainstream topics or patterns from the literature sample, the limitations or gaps in these existing studies, and the recommended future directions. The last step should be coherent with the bibliometric analysis, but also proceed beyond it. For example, the bibliometric analysis provides the quantitative measurement of the most highly cited scholars, journals, keywords, or institutions within the given domain. As a further step, the discussion could be more qualitative to offer insights for readers to understand what the quantitative measurements reflect and what could be performed in future research to fill the existing gaps.

Cases demonstrating bibliometric analysis in built environment

Following the workflow described in Figure 10.1 and the general principles of conducting bibliometric analysis in reviewing the defined topic in a given domain, the aim of this section is to extend the generalisation by demonstrating two cases in the built environment. Although the software tool adopted to conduct the bibliometric analysis was VOSviewer as the showcase, the option of software tools should not be limited to those presented in Table 10.1.

Case 1: Building information modelling and off-site construction

Following the review work conducted by Zhao (2017) who focused on BIM and Hosseini *et al.* (2018) targeting OSC, the aim of the review work in this case was to extend these two studies by investigating the existing literature about adopting BIM for OSC or BIM linked to OSC. Therefore, the scope of the review-based study was defined as BIM crossing over OSC. Examples of existing studies could be, but are not limited to, case studies of applying BIM to assist the early stage cross-disciplinary co-ordination in OSC projects; BIM in the life-cycle assessment of OSC projects from design and construction to facility management; and investigating the effects of BIM on OSC project delivery and productivity. Scopus was used as the database for the initial literature search. Other databases such as WOS could also be adopted.

The keyword search between BIM and OSC in Scopus is shown below.

> **TITLE-ABS-KEY** (BIM OR "Building Information Modelling" OR "Building InformationModeling") AND **TITLE-ABS-KEY** ("Off-site construction" OR "off site construction" OR "prefabricated construction" OR "industrialized building" OR "panelized construction" OR "modular construction" OR "tilt up construction" OR "offsite construction" OR "precast construction")

The reason for using "OR" relationships for BIM and OSC was that both terms could be presented in various phrases. There are a variety of different terminologies to express OSC, for instance, "modular construction". The "AND" relationship was used between the two major types of terms, meaning that in either the title, abstract, or keyword list, only publications with both terms mentioned would be selected in the automatic search of literature. Up until mid-February 2020, a total of 142 documents were found from Scopus. It can be seen that "BIM crossing over OSC" resulted in a much smaller literature sample compared with literature on BIM only or OSC. A total of 58 out of the whole initial sample were journal articles, with a majority of the rest of the literature coming from conference papers. Researchers must decide initially either to include the whole sample, regardless of document type, or to include journal articles only. Assuming the decision initially was to include only the 58 journal articles, the selected 58 journal articles could be saved in a single comma-separated values (CSV) file, including the information metrics, such as author, citation count, abstract, author keywords, and others. Scopus also seamlessly enables the export of literature information into other reference management tools such as Endnote. Indeed, similar procedures can be performed if WOS is adopted as the database instead of using Scopus, although the different coding format in WOS should be noted. Specifically, the format of **TITLE-ABS-KEY** shown in Scopus will be different in WOS, which allows mainly the search of the *Topic* and *Title* of documents.

Next, researchers must read through the downloaded CSV file, especially the title, abstract, and keyword list of each document, in order to perform the filtering and screening sub-steps illustrated in Figure 10.1. When the literature sample is larger (e.g. over 1,000 documents), it is essential that manual screening is performed by individual researchers. A comparison of the results is then made until final agreement is reached. It is common to exclude more documents after the manual reading and screening. In this case, a few documents selected initially were excluded because they fell without the pre-defined scope, for example, studies that were focused only on a technical aspect of BIM but without the focus on OSC and studies that were focused on OSC in which the role of BIM was not emphasised. After the literature sample had been finalised, the literature information contained in the CSV file could then be loaded into VOSviewer to conduct the networking analysis based on visualisation and citation-related metrics. The networking and citation analysis could be author keywords, as suggested by Oraee *et al.* (2017) and Hosseini *et al.* (2018), or other options, including journals, authors, institutions, and countries.

In this study, it was found that owing to the relatively small sample size of literature, the visualisation of keyword network could not display the links between keywords or the closeness of clusters of keywords being co-studied in the same articles. The relatively small sample size (less than 60 documents in this case) would motivate using other review methods, such as the content analysis conducted in Jin *et al.* (2019b) and the qualitative analysis based on manual review. The alternative approach would be to enlarge the sample size by including other types of literature, such as conference proceedings, and restart the process of bibliometric analysis.

Case 2: BIM and geographic information system

In this example, BIM is co-studied with the other digital tool, named geographic information system (GIS) in the built environment. Following a consistent procedure according to Figure 10.1, and using Scopus as the database, the following keywords relating to BIM and GIS were input to generate the initial literature sample.

> **TITLE-ABS-KEY** (BIM OR "building information modelling" OR "building information modeling") AND **TITLE-ABS-KEY** (GIS OR "geographic information system" OR "geographic information systems")

Continuing to follow consistently the workflow described in the former case (i.e. BIM and OSC review) and Figure 10.1 until the literature sample was finalised, researchers could now analyse the author keyword in VOSviewer or other software tools. Figure 10.2 illustrates the generated visualisation of keywords contained in the literature sample. The visualisation map in Figure 10.2 consists of the following elements: node with corresponding keyword texts, connection lines, and colour indicating the cluster. These elements were generated based on the in-built algorithms in VOSviewer by extracting the literature information (van Eck and Waltman, 2014). The example shown in Figure 10.2 indicates that inter-operability and IFC are two main frequently studied topics in linking BIM

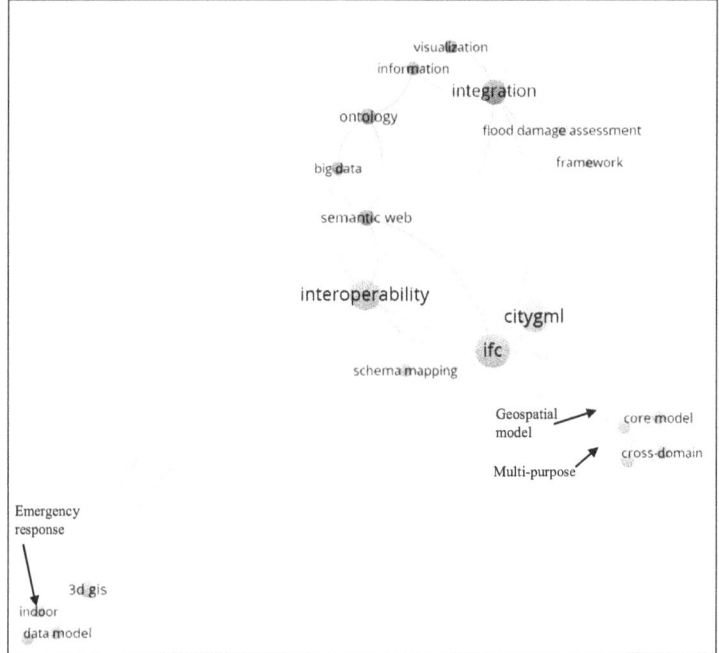

Figure 10.2 Visualised mapping of author keywords in VOSviewer.

to GIS. The clusters shown in Figure 10.2 were determined based on a given group of keywords being co-studied in the same publication, or one keyword being cited by another in different publications. The relevance of a pair of keywords, i.e. being co-studied within the same publication, can be detected in VOSviewer and visualised by using the connecting line in Figure 10.1. The distance between any two keywords shows how closely they are related. For example, inter-operability is the keyword that is closely related to IFC, which is a data format when exchanging information between BIM and GIS. Researchers are strongly recommended to include the critical analysis of clusters or the connections between keywords, for example, why and how are inter-operability, IFC, and schema mapping strongly connected in the existing studies, and what research has been performed under each cluster shown in Figure 10.2.

It should be noted that although VOSviewer was adopted as the showcase to demonstrate the usefulness of bibliometric analysis, the text mining and visualisation features of the bibliometric tools can be found in other software packages listed in Table 10.1. Besides the visualised map demonstrated in Figure 10.2, the quantitative measurements of author keywords can be summarised as presented in Table 10.2.

A variety of quantitative measurements are shown in Table 10.2, including the occurrence, which shows the popularity or the frequency of being studied; the average publication year, which depicts the recency of the keyword; and the

Table 10.2 Quantitative measurements of keywords generated from VOSviewer

Keyword	Occurrence	Average publication year	Average citations	Average normalised citations
IFC	11	2016	16.09	1.45
CityGML	9	2016	12.78	0.98
Inter-operability	9	2015	29.22	1.31
Integration	6	2017	13.83	1.60
3D GIS	3	2014	37.67	0.97
Ontology	3	2016	13.67	1.04
Semantic Web	3	2017	20.67	2.25
Big Data	2	2017	3.00	0.33
Core Model	2	2018	0.00	0.00
Cross-Domain	2	2018	0.00	0.00
Data Model	2	2014	49.00	1.32
Emergency Response	2	2015	25.00	1.28
Flood Damage Assessment	2	2016	15.50	1.71
Framework	2	2016	13.50	1.49
Geospatial Model	2	2018	0.00	0.00
Indoor	2	2014	49.00	1.32
Information	2	2016	16.00	1.90
Multi-Purpose	2	2018	0.00	0.00
Schema Mapping	2	2016	24.50	2.34
Visualisation	2	2015	16.50	1.39

Source: Original.

citation-related indicators, which measure the influence of the given keyword. There is a total citation number for any given keyword listed in Table 10.2 and also for any given journal, author, institution, or country. Average citation is calculated automatically in the software tool by dividing the total citation by the number of documents. The normalised citation, or average normalised citation, is the measurement that corrects the bias by which older documents or keywords accumulate more time to receive citations. Normalised citation is calculated by dividing the total citations (or the average citation of the given keyword in this case) by the average number of citations of this keyword in the same year. Generally, the normalised or average normalised citation can be considered to be the unbiased measurement of the influence of a given keyword, linked journal, author, institution, or country of source within the defined domain. More detailed descriptions of these quantitative indicators can be found in Jin *et al.* (2019a). Overall, it is noted that the bibliometric analysis is not the last step of the science mapping methodology according to the workflow described in Figure 10.1. Researchers in this study of BIM linked to GIS are encouraged to take a step further to perform the in-depth discussion, especially when addressing the three research questions relating to the mainstream topics in the defined scope, gaps, or limitations, and the future research directions. It would be necessary for researchers to have their own understanding, reflection, and critical thinking about the bibliometric outputs to address the three generalised research questions within the specifically defined scope (e.g. BIM integration with GIS in this case).

Conclusions

In this chapter, bibliometric analysis was introduced as one research method to form part of the science mapping methodology and the extent to which this tool can assist in the review of secondary sources, such as academic publications, was discussed. The chapter started from the generalised concept and steps of conducting bibliometric analysis. A general guide for conducting the bibliometric analysis was provided, including secondary data sources, and the commonly adopted software tools, based on text mining and visualisation features. Informed judgement considering the suitability of conducting a bibliometric analysis was given. Examples included cases with limited sample size of literature (e.g. less than 50 documents identified). The size of the available literature sample is directly related to the defined topic from being broader or narrower in scope.

Some further recommendations were provided in the chapter. First, bibliometric analysis does not exclude other research methods in conducting a review of secondary databases. Indeed, there is a possibility of linking bibliometric analysis with other review methods such as content analysis. The review scope of the topic to be studied could vary depending on the aim of the research. It is not uncommon that within the same study, the review is started with a wider scope, with a larger literature sample, and then is narrowed down to a focal point. Second, not dissimilar to other review methods, it is necessary for researchers in bibliometric analysis to decide on the keyword and database for the literature search, and also to decide on the type of literature (e.g. conference proceedings) to be included. The selection of keywords reflects the review scope, which together with literature type determines the literature sample size. Third, bibliometric analysis is one of the methods or steps involved in achieving a literature review, but it cannot replace the further in-depth discussion. The aim of the further discussion is to address three general research questions, namely: (1) what are the mainstream research keywords or topics within the pre-defined scope?; (2) what are the limitations or gaps in these existing studies?; and (3) what are the recommended future research directions? It is key to maintain the coherence of the discussion linked to the prior step of bibliometric analysis. To demonstrate these main recommendations, two examples of conducting the science mapping approach were showcased. In the first example of literature review, the scope was focused on BIM applied in OSC. A limited literature sample was found as only journal articles were included to conduct the bibliometric analysis. Therefore it was suggested that other review methods be adopted, such as content analysis or alternatively to include other document types (e.g. conference proceedings). The second showcase was used to demonstrate the review of BIM working with GIS in the built environment. Author keyword was studied by adopting the bibliometric analysis through visualisation and quantitative measurements, such as frequency to measure the popularity of a given keyword and the citation-related indicators to measure the influence of the studied keyword.

Finally, various misconceptions or mistakes should be avoided in conducting a bibliometric analysis. The first mistake is to treat the bibliometric analysis as the aim rather than the methodology. The aim should not be simply to generate some eye-catching visualisation maps. Instead, bibliometric analysis is a method or an approach to achieve the pre-defined research objectives. It should not be treated as the aim of the research. In other words, the bibliometric analysis is a tool to assist in achieving the research goal or in answering the research questions by searching mainstream keywords or topics in the pre-defined domain. Researchers are encouraged to take a step further beyond the network analysis of the citation data (e.g. keyword, document) by demystifying the selected research domain. The second mistake is applying the bibliometric analysis in a field or domain with which the researcher is not familiar. The researchers should use their own critical thinking to interpret the visualisation maps of keywords and to address further the research questions relating to the limitations of existing studies and the recommended future research directions. A third mistake is to perceive the bibliometric analysis as a fully automatic method to be used in screening the literature sample. The science mapping workflow described in this chapter does not allow Artificial Intelligence to filter automatically and screen the initially selected literature from the database. Before conducting the bibliometric analysis, manually reading the abstract, title, or author keywords is still required to exclude further documents which do not fall within the defined scope of the research.

References

Aghaei Chadegani, A., Salehi, H., Md Yunus, M.M., Farhadi, H., Fooladi, M., Farhadi, M. and Ale Ebrahim, N. (2013). A comparison between two main academic literature collections: Web of Science and Scopus databases. *Asian Social Sciences*, 9(5), pp. 18–26.

Aria, M. and Cuccurullo, C. (2017). Bibliometrix: An R-tool for comprehensive science mapping analysis. *Journal of Informetrics*, 11(2017), pp. 959–975.

Bastian, M., Heymann, S. and Jacomy, M. (2009). Gephi: An open source software for exploring and manipulating networks. *International AAAI Conference on Weblogs and Social Media*, San Jose, CA, pp. 361–362.

Brookes, B. C. (1990). Biblio-, Sciento-, Infor-metrics??? What are we talking about? In: L. Egghe, R. Rousseau (Eds), Informetrics 89/90. Selection of Papers Submitted for the Second International Conference on Bibliometrics, Scientometrics and Informetrics, Amsterdam, Netherlands, Elsevier, pp. 31–43.

Burghardt, K.J., Howlett, B.H., Fern, S.M. and Burghardt, P.R. (2020). A bibliometric analysis of the top 50 NIH-Funded colleges of pharmacy using two databases. *Research in Social and Administrative Pharmacy*. In press. doi:10.1016/j.sapharm.2019.10.006.

Butler, L. and Visser, M.S. (2006). Extending citation analysis to non-source items. *Scientometrics*, 66(2), pp. 327–343.

Chen, C. (2016). CiteSpace: Visualizing patterns and trends in scientific literature. Available at: http://cluster.cis.drexel.edu/~cchen/citespace/ [Accessed on: 6 March 2019].

Chen, W., Jin, R., Xu, Y., Wanatowski, D., Li, B., Yan, L., Pang, Z. and Yang, Y. (2019). Adopting recycled aggregates as sustainable construction materials: A review of the scientific literature. *Construction and Building Materials*, 218, pp. 483–496.

He, Q., Wang, G., Luo, L., Shi, Q., Xie, J. and Meng, X. (2017). Mapping the managerial areas of building information modeling (BIM) using scientometric analysis. *International Journal of Project Management*, 35(4), pp. 670–685.

Hood, W. and Wilson, C. (2001). The literature of bibliometrics, scientometrics, and informetrics. *Scientometrics*, 52(2), pp. 291–314.

Hosseini, M.R., Martek, I., Zavadskas, E.K., Aibinu, A.A., Arashpour, M. and Chileshe, N. (2018). Critical evaluation of off-site construction research: A scientometric analysis. *Automation in Construction*, 87, pp. 235–247.

Jiang, W., Martek, I., Hosseini, M.R. and Chen, C. (2019). Political risk management of foreign direct investment in infrastructure projects: Bibliometric-qualitative analyses of research in developing countries. *Engineering, Construction and Architectural Management*. In Press. doi:10.1108/ECAM-05-2019-0270.

Jin, R., Gao, S., Cheshmehzangi, A. and Aboagye-Nimo, E. (2018). A holistic review of off-site construction literature published between 2008 and 2018. *Journal of Cleaner Production*, 202, pp. 1202–1219.

Jin, R., Yuan, H. and Chen, Q. (2019a). Science mapping approach to assisting the review of construction and demolition waste management research published between 2009 and 2018. *Resource, Conservation and Recycling*, 140, 175–188.

Jin, R., Zhong, B., Ma, L., Hashemi, A. and Ding, L. (2019b). Building information modelling for building performance analysis in the design life-cycle. *Automation in Construction*, 106, pp. 102861.

Oraee, M., Hosseini, M.R., Papadonikolaki, E., Palliyaguru, R. and Arashpour, M. (2017). Collaboration in BIM-based construction networks, a bibliometric-qualitative literature review. *International Journal of Project Management*, 35(7), pp. 1288–1301.

Qiu, J., Zhao, R., Yang, S. and Dong, K. (2017). *Informetrics: Theory, Methods and Applications*, Springer, Singapore.

Rousseau, D.M. (ed.) (2012). *The Oxford Handbook of Evidence-Based Management*, Oxford University Press, Oxford, UK.

Song, J., Zhang, H. and Dong, W. (2016). A review of emerging trends in global PPP research, analysis and visualization. *Scientometrics*, 107(3), pp. 1111–1147.

Van Eck, N.J. and Waltman, L. (2010). Software survey: VOSviewer, a computer program for bibliometric mapping. *Scientometrics*, 84(2), pp. 523–538.

Van Eck, N.J. and Waltman, L. (2014). Visualizing bibliometric networks. In: Y. Ding, R. Rousseau and D. Wolfram (eds.), *Measuring Scholarly Impact*, Springer, Cham, pp. 285–320.

Vilutiene, T., Kalibatiene, D., Hosseini, M.R., Pellicer, E. and Zavadskas, E.K. (2019). Building information modeling (BIM) for structural engineering: A bibliometric analysis of the literature. *Advances in Civil Engineering*, 2019, Art. 5290690.

Wang, L., Huang, M., Zhang, X., Jin R. and Yang T. (2020). A review of BIM adoption in the higher education of AEC disciplines. *Journal of Civil Engineering Education*, 146(3), p. 06020001.

Wang, C. and Pang, X. (1998). Terms of bibliometrics (1). *Information Studies: Theory & Application*, 21(1), pp. 61–61.

Xu, Y., Zeng, J., Chen, W., Jin, R., Li, B. and Pan, Z. (2018). A holistic review of cement composites reinforced with graphene oxide. *Construction and Building Materials*, 171(2018), pp. 291–302.

Yang, S. and Yuan, Q. (2017). Are scientometrics, informetrics, and bibliometrics different? *Conference: the 16th International Conference for Scientometrics & Informetrics*, 16–20 October 2017, Wuhan, China, ISSI, pp. 1507–1518.

Zhao, X. (2017). A scientometric review of global BIM research: Analysis and visualization. *Automation in Construction*, 80, pp. 37–47.

Zhao, X., Ke, Y., Zuo, J., Xiong, W. and Wu, P. (2020). Evaluation of sustainable transport research in 2000–2019. *Journal of Cleaner Production*, 256, p. 120404.

11 Scientometric review and analysis

A case example of smart buildings and smart cities

*Timothy O. Olawumi, Abdullahi B. Saka,
Daniel W.M. Chan, and Nimesha S. Jayasena*

Introduction

Understanding the dynamics of knowledge in the various disciplines is vital not only to expanding the knowledge base but also to identifying the diverse aspects of such disciplines. Research techniques, such as scientometrics, bibliometrics, and informetrics, provide avenues through which to study and reflect on the dynamics of a discipline. There are significant overlaps between these three techniques in terms of their methodologies, theories, and applications, but they differ in their subject background (Mooghali *et al.*, 2011). Bibliometrics was designed to analyse books and articles statistically and other forms of communication, while scientometric, as its name implies, is focused on scientific publications (otherwise known as the science of science) with the motive being to guide decision-making or policy formulation. However, informetrics has been streamlined for the domain of information science and, thus, has found limited application across disciplines. Brookes (1990) and Hood and Wilson (2001) provided a further in-depth discussion on the history, inter-relationships, and differences between these three statistical record techniques. An application of bibliometric research in the built environment can be seen in the study of Olawumi *et al.* (2017). Owing to word limitations, readers interested in the similarities and differences of the scientometric, bibliometric, and informetric analyses can refer to Mooghali *et al.* (2011) and Qiu *et al.* (2017).

Scientometric analysis or review has several definitions in the literature. Nalimov first coined the term in 1969 (Siluo and Qingli, 2017). Olawumi and Chan (2018) described the scientometric analysis as a technique that enables "concise capturing and mapping of scientific knowledge", while according to Qiu *et al.* (2017), it is a discipline which employs statistical methods to "quantify the scientific research personnel and their achievements". Also, according to Tague-Sutcliffe (1992), scientometrics is the quantitative scrutinising of scientific activities such as publication records. Chen and Song (2019) define it as a "research of literature-based discovery". In recent years, scientific mapping software, such as CiteSpace, VOSviewer, Gephi, and BibExcel, among others, has been adopted to generate a visualisation and overview of the underlying knowledge dynamics.

The aim of the current study, as discussed in this chapter, was to illustrate and present scientometric network analysis as a secondary research methodology, using the CiteSpace software. A case study, on the theme smart buildings and smart cities, was adopted in this chapter to show the application of scientometric analysis as a secondary research methodology in the built environment. The rationale of the study was to provide an in-depth guide to readers in the use of scientometric analysis of the literature towards enabling them to map the trend and structure of any research field or topic. The scientometric analysis of the smart building and cities research field is used in this chapter to track the evolution of the concepts, establish the trending research themes, and identify the key research clusters. The study is expected to guide new entrants to this field towards becoming well-established researchers seeking collaboration opportunities.

Smart buildings and smart cities

During the last two decades, the concept of smart cities has become a widespread theme of discussion in scientific literature and international policies (Albino *et al.*, 2015). Rana *et al.* (2019) defined a smart city as "a technologically advanced and modernised territory with a certain intellectual ability that deals with various social, technical, economic aspects of growth based on smart computing techniques to develop superior infrastructure constituents and services" (Rana *et al.*, 2019, p. 503). As described by Bakıcı *et al.* (2013), a smart city interconnects people, information, and city elements in order to create a sustainable city. Cities are becoming more complex every day with the rising expectations of the characteristics in modern cities together with rapid urbanisation (Nam and Pardo, 2011). According to Peris-Ortiz *et al.* (2017), rapid urbanisation results in complex challenges to managing cities in terms of achieving sustainable urban development. These challenges have anticipated the requirement of smart cities and escalated the development of strategies to enable the realisation of smart cities. Schaffers *et al.* (2011) suggested that the concept of a smart city is a response to the requirement to guide pathways of urban development in strategic directions to address the challenges and achieve sustainability.

With reference to Lazaroiu and Roscia (2012) and Bakıcı *et al.* (2013), a smart city represents a society, which consists of average technological capacity, interconnectedness, sustainability, comfortability, attractiveness, and security (Bakıcı *et al.*, 2013). The development of smart cities has gained widespread attention in research, practice, and policies based on the belief that smart cities create a more liveable environment, which will provide more benefits for the citizens (Milenković *et al.*, 2017). According to Ramaprasad *et al.* (2017, p. 15), the concept of smart city was identified as a "multi-disciplinary concept that embodies not only its information technology infrastructure but also its capacity to manage the information and resources to improve the quality of lives of its people".

Since most of our lives are spent in buildings and/or using built infrastructure, smart buildings will constitute necessarily a critical component of smart city development. Kathiravelu *et al.* (2015) defined smart buildings as a scenario of the prevalent use of ubiquitous computing, integrating IoT elements, including sensors, computing elements, and control algorithms incorporated into the buildings. Smart buildings differ from conventional buildings from inception of the designing process, and have wider potential and benefits than merely remote control (Batov, 2015). Chourabi *et al.* (2012) and Soyinka *et al.* (2016) highlighted the importance of smart buildings in achieving sustainable urban development to overcome the current urban challenges. Consequently, there has been an increase in research output on smart buildings and cities in the built environment. Hence, the purpose of this chapter was to map these studies towards providing readers with an in-depth understanding of the key issues in research themes concerning smart buildings and cities.

Usefulness of and approaches to scientometric analysis

The use of scientometric reviews plays a key role in synthesising structural patterns, identifying the direction and frontiers of research, extracting original findings from publications, and assessing the performance of authors and institutions, among others, within the pre-defined research field. Chen and Song (2019) suggested that scientometric review could help to identify challenges and difficulties being faced in the evolution of a scientific field. More so, according to Chen and Song (2019), it could be a valuable tool for early researchers to identify saturated and emerging research themes towards providing them with an overview and visualisation of the intellectual landscape of the research field. A scientometric review can also help in the characterisation of the development of a research field (Mooghali *et al.*, 2011) and the mapping of the various research clusters (Olawumi and Chan, 2018). With the advent of scientometric software, which only makes use of the records of publications, such as title, abstract, keywords, acknowledgement, and references, without the main body of the research publication, mapping a research field might be slightly or more disadvantaged using the available mapping software.

There are three main approaches to using scientometrics to analyse a specific research field: the influence metric, the intellectual composition, and the knowledge base metric (Siluo and Qingli, 2017; Olawumi and Chan, 2018). The influence metric is focused on measuring the influence and co-operation among authors, using criteria such as their institutional and geographical affiliations, publishing journals, languages, document types, and research funding. The intellectual composition metric is used to examine and address the development and evolution of the research field by taking into consideration aspects, such as research keywords, subject areas or categories, research clusters, and methodological approaches. The emphasis of the knowledge base metric is on measuring the longitudinal distribution of the research growth and citations, visualising h-index analysis and geospatial analysis as well as the emerging, salient, and future direction of the pre-defined research field.

Research method

This section contains an overview of a typical research approach for scientometric analysis.

Defining the research problem

This involves defining the purpose of the study and search technique, which are related and vital to the overall quality of the study. It is an important task because the quality of the output depends on the input, and the result of the analysis would depend on it. The most common technique is the keyword search technique. The keywords which serve as the query should be chosen carefully to reflect the research domain, and should be reviewed by domain experts. This process is non-trivial as the keywords should be refined iteratively before final adoption (Chen and Song, 2019). Chen and Song (2019) proposed a cascading, citation expansion search technique by backward or forward expansion from a seed article. However, the method requires "constant programmatic access to a master source of scientific articles" (Chen and Song, 2019). A search query of "smart buildings" or "smart cities" was used in this study.

Data retrieval

There are two different types of databases, which are the citation databases and bibliographic databases. The citation databases are more comprehensive and detailed as they contain both bibliographical and citation information (Jayasree and Baby, 2019). Data for scientometric analysis are retrieved often from citation databases, and those used most in the built environment are Scopus and Web of Science (WoS). Other databases include Google Scholar, Dimension, CiteseerX, Pubmed, and MathSciNet. The decision about which of the databases to use depends often on the purpose of the study. WoS is a database that contains more influential journals, while Scopus has a broader coverage compared with WoS (Saka and Chan, 2019a). Combining different databases is encouraged to cover as many datasets as possible. However, the major challenge is the removal of repetitive data and dataset forms. Thus, Scopus, WoS, or other databases are being used separately as a source of data for analysis. WoS was adopted as the database in this study with a search query of "smart buildings" or "smart cities", which resulted in 28,962 documents.

Pre-processing

The output of the data search should be refined to suit the aim of the study. This may include refining according to the document type, language, year, countries/regions, and research area. The document type includes articles, conference proceedings, book chapters, and other materials. Articles are adopted often because they contain the latest developments in the research domain. However, all the document types can be combined for various reasons, such as new research areas, and when the aim of the study is to evaluate holistically or to avoid publication bias (Saka and Chan, 2019b). Also, depending on the aim of the study, some countries/years/languages/research areas can be excluded. It is noteworthy that the refining options often serve as a limitation and might include bias in the study. Thus, refining the data search should be considered diligently.

In this study, the output was refined using built environment research areas, articles (as document types), English language, and year range from 2005 to 2019. The year 2020 was not included because more articles would be published, and a minimum span of 10 years is sufficient to show intellectual evolution in a research domain (Jin *et al.*, 2018). The pre-processing stage resulted in 1,564 journal articles that served as the input dataset in this study.

Data analysis tools

Many tools are used for scientometric analysis, including CiteSpace, VOSviewer, CitNetExplorer, Sci^2, BibExcel, HistCite, Pajek, Publish or Perish, Scholarmeter, and Gephi (Jayasree and Baby, 2019). These tools have strengths and weaknesses, and there are no one fits all tools. However, the most popular tools in the built environment are CiteSpace, VOSviewer, and Gephi (Oraee *et al.*, 2017; Hosseini *et al.*, 2018; Darko *et al.*, 2020).

1 CiteSpace is a free Java application created by Chaomei Chen for visualising and analysing the intellectual evolution of research domains. The data source for CiteSpace includes WoS, Scopus, Lens, CSCD, CSSCI, and PubMed. The application uses both a time-based and graphical approach for visualisation. The various nodes that can be generated with the tool include a co-authorship network, network of co-authors' institutions, network of co-authors' countries, the network of co-occurring phrases, document co-citation network, author co-citation network, and journal co-citation network, among others. Refer to Chen (2014) for further details about the use of CiteSpace. The tool provides comprehensive analysis options, but this might be overwhelming for new users.

2 VOSviewer was created by Nees Jan Van Eck and Ludo Waltman. It uses distance-based visualisation of the network, and the software is easy to use but offers less functionality compared with CiteSpace (Van Eck and Waltman, 2014). The major functionalities of the tool are to create maps and to explore the maps with input from databases such as WoS, Scopus, Dimensions, PubMed files, and reference files such as RIS, RefWorks, and Endnote files. Refer to van Eck and Waltman (2019) for further details about use.

3 Gephi is focused more on network visualisation than network analysis (van Eck and Waltman, 2014). Gephi is a software application for visualising, manipulating, exporting, spatialising, and filtering all types of networks (Bastian *et al.*, 2009). Thus, it is often combined with other tools for analysis.

The combination of CiteSpace and VOSviewer (Saka and Chan, 2019a) or the combination of CiteSpace, VOSviewer, and Gephi (Darko *et al.*, 2020) is becoming more popular in the built environment. CiteSpace was adopted in this study for visualisation of the smart building and smart cities research domain.

Data analysis techniques

The following are some of the conventional analysis techniques using CiteSpace:

a   Co-author analysis:

   a   Co-authorship network: The network presents the relationship between the authors in the dataset. Nodes represent the authors and the links represent the collaboration between the authors. This network shows the porosity of the research domain and how the researchers collaborate and interact with each other to form smaller research communities.

   b   Network of institutions/faculties and countries/regions: The network presents the contributions of institutions and countries to a research domain and the collaboration between them.

b   Co-word analysis

   a   Network of co-occurring keywords: Keywords are essential parts of research publications, be they journals, conference papers, books, magazines, and even webs or blogs (where they are referred to as "tags"). They are assigned to a piece of information for better description and indexing. According to Olawumi and Chan (2018), keywords provide a more concise way to understand a concept as well as the content of research publications. According to Zhao (2017), keywords illustrate the trend of a research field. In research publication, keywords are categorised broadly in two ways: author keywords and keyword plus (Olawumi and Chan, 2018). Author keywords, as the name infers, are keywords provided by the authors of the publication, while keyword plus is based on the classification of the publishing journal.

   b   Network of co-occurring subject categories: This network evaluates the subject categories of the documents in the dataset. The subject categories are usually assigned by the database, depending on the scope of the document.

c   Co-citation analysis

   a   Author co-citation network: The author co-citation network provides a pattern of connection among the diverse authors whose research publication appears as cited references within the same journal paper (Olawumi and Chan, 2018).

   b   Document co-citation network: The network evaluates the cited references in the dataset to show the articles/documents that have been highly cited and referred in the dataset.

   c   Journal co-citation: The network presents the co-cited journals in the dataset and inferences can be drawn from the aim and scope of the top-cited journals as regards the research direction in the dataset.

d   Citation burst and centrality: The citation burst analysis within the CiteSpace software application is based on Kleinberg's algorithm (Kleinberg, 2002), and it portrays the citation increase within a short period (Olawumi and Chan, 2018). Meanwhile, the betweenness centrality is based on the work of Free-man (1977), and is described as the degree to which a node or point on the network lies within the shortest path between other nodes.

Data analysis and results

CiteSpace was used to generate the networks, presented below, using the 1,564 articles from WoS as the input. Since the period examined in the study was from the year 2005 to 2019, the "Years per slice" option was set to 1, and selection criteria were set to top 20 levels of the most cited or occurred items from each slice. Also, the pathfinder utility option in CiteSpace was used for the network pruning to remove redundant links.

Co-author network

The network shows the collaboration between the authors in the 1,564 articles with 194 nodes and 184 links. The network has modularity (Q) of 0.967 and mean silhouette (S) of 0.625, which are the quantitative representation of the network structure. The Q relates to the overall structural properties of the network, and a value greater than 0.7 reflects the porosity of the clusters, while the S indicates the homogeneity of the clusters (Chen, 2014). Thus, the network consists of loosely packed clusters that are homogenous, as shown in Figure 11.1. The size of the nodes corresponds to the number of author publications in the dataset, and authors such as Satish Nagarajaiah and Billie F. Spencer, Jr. are noticeable in the network. The top five productive authors in the network are shown in Table 11.1 (Section A). Most of the research clusters are smaller in size, which indicates that they do not collaborate significantly with other clusters. This shows the porosity of the research domain of "smart building" or "smart cities". Citation burst occurred in 2005 for Billie F. Spencer, Jr. (strength = 3.27, 2005–2006) and Shankar Narasimhan (strength = 3.27, 2006–2008). This means that the work of the authors gained significant attention from researchers in this domain during the specified period. This often coincides with the publication of notable or significant research works that attract more citations from other researchers.

Analysis of co-occurring keywords

The network analysis of the co-occurring keywords, as shown in Figure 11.2, has 136 nodes and 388 links. According to Olawumi and Chan (2018), the node sizes represent the frequency of occurrence of the keyword in the dataset. The keyword analysis network also has modularity, Q = 0.5013, and a mean silhouette value, S = 0.7117. The Q-value implies that the nodes within the network are moderately packed, while the S-value shows a high homogeneity in the keyword clusters. The network analysis revealed some high-frequency keywords

Figure 11.1 Co-author network.

(Figure 11.2) in the research *corpus* which were: "system" (frequency, f = 229), "smart city" (f = 219), "model" (f = 144), "city" (f = 138), "building" (f = 129), "performance" (f = 125), "management" (f = 101), "design" (f = 88), "optimization" (f = 88), "sustainability" (f = 85), "simulation" (f = 80), "smart grid" (f = 80), and "energy" (f = 75).

The influence and significance of the keywords were analysed using the betweenness centrality and citation burst. Also, the centrality scores were

Table 11.1 Scientometric analysis for co-authors, keywords, and document co-citation network

1A – Top authors

Author	Institution	Country (counts)
Satish Nagarajaiah	Rice University	United States (7)
Billie F. Spencer, Jr.	University of Illinois at Urbana-Champaign	United States (5)
Shankar Narasimhan	Indian Institute of Technology Madras	India (5)
Rodney A. Stewart	Griffith University	Australia (4)
Jeong Tai Kim	Kyung Hee University	South Korea (4)

1B – Keywords citation bursts

Keywords	Burst strength	Span
Structural control	10.2232	2006–2015
Demand response	9.3014	2014–2017
Technology	8.0586	2017–2019
Smart grid	7.8353	2015–2016
Internet of thing	7.7321	2017–2019
Bridge	6.5231	2006–2011
Active control	5.8332	2008–2010
Energy consumption	5.6142	2014–2017
Neural network	4.8727	2006–2015
Hybrid control	3.7626	2006–2009
Identification	3.6335	2006–2010
Smart structure	3.5736	2008–2015
Policy	3.4641	2011–2015
Prediction	3.4492	2014–2015

1C – Document citation and betweenness centrality

Article	Centrality	Total citation
Caragliu *et al.* (2011)	0.21	50
Neirotti *et al.* (2014)	0.09	50
Albino *et al.* (2015)	0.03	45
Ahvenniemi *et al.* (2017)	0.01	33
Batty *et al.* (2012)	0.43	32

Source: Original.

normalised between the interval of 0 and 1, and a node with a higher centrality score links two or more large clusters of nodes (Chen, 2014; Olawumi and Chan, 2018). Such nodes also help to pinpoint key and critical research publications.

Keyword nodes with betweenness centrality scores included: "building" (centrality = 0.36), "system" (0.25), "performance" (0.25), "structural control" (0.21), "design" (0.19), "model" (0.17), and "optimization" (0.17), among others. These keyword themes were shown to be shaping and connecting the development of

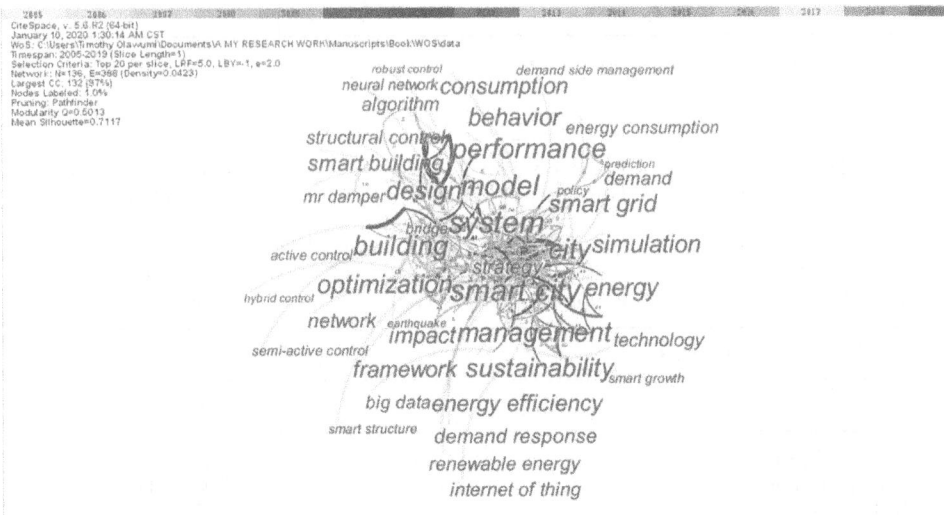

Figure 11.2 Network analysis of co-occurring keywords.

the emerging concept of smart buildings and smart cities. For the co-occurring keywords citation burst, 14 keywords were identified from the analysis network. These keywords, with citation burst, as shown in Table 11.1 (Section B), were the salient topics and themes relating to smart buildings and smart cities. A finding of interest was that keywords, such as "structural control", "demand response", "technology", "smart grid", "internet of things", "energy consumption", and "neural network" had citation bursts and high frequencies. The results portend that these salient research themes are critical to developing the smart buildings and cities within the built environment.

Analysis of author co-citation network

The research *corpus* extracted from the WoS records formed the dataset for the author co-citation analysis, as shown in Figure 11.3, which has 299 nodes and 1,142 links. The network had modularity (Q = 0.7449) and a mean silhouette (S = 0.6089), which showed slightly loose clusters of authors. Also, the node size of each author in the analysis network indicates its co-citation frequency, while the links show an "indirect co-operative alliance" between the authors based on the metric of their co-citation frequency. Based on the network analysis (Figure 11.3), the ten highly cited authors were identified, of which two of the most cited authors were international organisations, which reflected significant interest in the concept of smart buildings and cities worldwide. These highly cited authors were the European Commission (frequency, f = 74, Belgium*), Spencer Billie (f = 65, United States), Caragliu Andrea (f = 63, Italy), Giffinger Rudolf (f = 58, Austria),

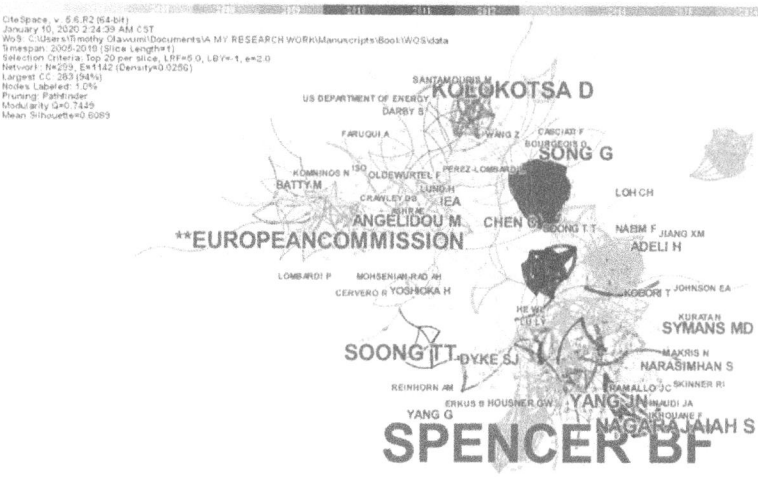

Figure 11.3 Network analysis of authors' co-citation.

Yang Jann (f = 50, United States), the UN (f = 45, United States*), Batty Michael (f = 45, United Kingdom), Nagarajaiah Satish (f = 44, United States), Neirotti Paolo (f = 44, Italy), and Komninos Nicos (f = 42, Greece). The diversity in the affiliation of the authors showed the growing interest and evolution of the research fields of smart buildings and cities.

The authors, with high citation bursts within a short period, were identified from the analysis network (Figure 11.3). These authors included Spencer Billie (burst strength = 15.65, 2005–2014), Yang Jann (burst strength = 13.08, 2005–2010), Nagarajaiah Satish (burst strength = 12.29, 2006–2010), Narasimhan Sriram (burst strength = 12.16, 2006–2010), Lombardi Patrizia (burst strength = 9.56, 2017–2019), Perez-Lombard Luis (burst strength = 8.97, 2016–2019), Dyke Shirley (burst strength = 8.31, 2006–2014), Angelidou Margarita (burst strength = 8.29, 2017–2019), and Dong Bing (burst strength = 8.18, 2017–2019). Publications, including communique and research papers by these authors, have shaped the concept and research field of smart buildings and smart cities. Hence, their works are worth following.

Nodes, with betweenness centrality, and their values were identified from the analysis network which were Deb Kaushik (centrality = 0.28), European Commission (0.26), Kolokotsa Denia (0.19), Song Gangbing (0.19), Soong Tsu Teh (0.18), Yang G (0.15), and Spencer Billie (0.13), among others. These authors had made notable and influential contributions in the research fields of smart buildings and smart cities. These authors also helped to connect the various research clusters and communities.

Document co-citation network

Figure 11.4 shows the document co-citation network with modularity of 0.83 and a mean silhouette of 0.41, which depicts a loosely clustered network but less homogenous than cases A and B. Table 11.1C shows the list of the top five documents that were well cited in the dataset. These documents were well placed in the network, as shown in Figure 11.4. Notably, Neirotti *et al.* (2014) examined the concept of smart city and assessed the trend at a global level. The network shows six categories, sub-categories, and the coverage index (CI) was defined. From the study it was evident that a unified concept for smart cities was lacking and it was concluded that the concept was contextual. Similarly, Caragliu *et al.* (2011) and Albino *et al.* (2015) also examined the concept of a smart city. This suggested that the concept of smart city is multi-faceted and dynamic.

Citation burst occurred for 18 of the articles and the top five included: Spencer Jr. and Nagarajaiah (2003) (burst strength = 9.05, 2006–2011), Ramallo *et al.* (2002) (burst strength = 8.15, 2006–2010), Zanella *et al.* (2014) (burst strength = 5.57, 2017–2019), Oldewurtel *et al.* (2012) (burst strength = 5.43, 2016–2017), Yang and Agrawal (2002) (burst strength = 4.78, 2006–2008). The most recent burst included Kramers *et al.* (2014) (burst strength = 3.96, 2017–2019) and Zanella *et al.* (2014) (burst strength = 5.57, 2017–2019). The top articles with citation burst were related more to smart buildings, which include smart structures, while the latest articles with burst related more to smart cities. It could be deduced that the concept of smart structures preceded the concept of smart cities, which has been gaining more attention in recent years.

Figure 11.4 Document co-citation network.

Conclusions

The use of scientometric analysis is becoming widespread because of the increase in research outputs. The use of scientometric analysis varies, depending on the aim of the study, which might be for comparison of the research domains, intellectual evolution of the research domains, or a combination of both. Consequently, the aim, data retrieval approach, pre-processing, analysis tools, and techniques are of utmost importance in the scientometric analysis, as these would determine the quality of the outputs.

A simplified application of scientometric analysis has been presented in this chapter for the research domain of smart buildings and smart cities as an example for further illustration. The method will continue to gain widespread usage in research because of its usefulness and the meteoric increase in research outputs over the years.

Although scientometric analysis is easy to use and apply to the *corpus* of articles, it requires a comprehensive understanding of the research domains. Also, the method can be used as a secondary research method in the built environment because of its rigour and quantitative justifications. In this chapter, articles not written in English were not part of the analysed *corpus*, which was a limitation to the study. Also, researchers interested in applying the scientometric analysis of research areas can follow the steps illustrated under the method section towards replicating the scientometric approach.

CiteSpace was used as a tool for the scientometric analysis of the trend and structure of smart buildings and cities in the extant literature via the generation of the co-author network, co-occurring keywords network, co-author citation network, and document co-citation network. Keywords such as "structural control," "demand response," "technology," "smart grid," "Internet of thing," and "energy consumption" were determined as the salient keywords with the highest burst strength and are the significantly important topics and themes in smart buildings and smart cities. Therefore, it can be determined that these keywords play an essential role in the development of the research areas of smart buildings and smart infrastructure.

More so, the chapter identified key researchers such as Deb Kaushik, Kolokotsa Denia, Spenser Billie, among others, who have contributed and are influential to the development of the concepts of smart building and cities. Hence, it is recommended for research students and researchers interested in the field of smart buildings and cities to follow their work. The diverse countries of the first ten highly cited authors, which include Belgium, United States, Italy, Austria, the United Kingdom, among others, illustrate growing interest in the research areas of smart buildings and cities. Hence, the highlights of the key authors, keywords, and research clusters provide relevant information for researchers who are interested in collaborations within the areas of smart buildings and cities. Future studies can work on the salient as well as the upcoming research themes identified in the study towards undertaking in-depth research on it.

References

Ahvenniemi, H., Huovila, A., Pinto-Seppä, I. and Airaksinen, M. (2017). What are the differences between sustainable and smart cities? *Cities*, 60, pp. 234–245. doi:10.1016/j.cities.2016.09.009.

Albino, V., Berardi, U. and Dangelico, R.M. (2015). Smart cities: Definitions, dimensions, performance, and initiatives. *Journal of Urban Technology*, 22(1), pp. 3–21. doi:10.1080/10630732.2014.942092.

Bakıcı, T., Almirall, E. and Wareham, J. (2013). A smart city initiative: The case of Barcelona. *Journal of the Knowledge Economy*, 4(2), pp. 135–148.

Bastian, M., Heymann, S. and Jacomy, M. (2009). Gephi: An open source software for exploring and manipulating networks. Paper presented at the Third International ICWSM Conference, May 17–20, San Jose, CA, 2 pages.

Batov, E.I. (2015). The distinctive features of "smart" buildings. *Procedia Engineering*, 111, pp. 103–107.

Batty, M., Axhausen, K.W., Giannotti, F., Pozdnoukhov, A., Bazzani, A., Wachowicz, M. and Portugali, Y. (2012). Smart cities of the future. *The European Physical Journal Special Topics*, 214(1), pp. 481–518. doi:10.1140/epjst/e2012-01703-3.

Brookes, B.C. (1990). Biblio-, sciento-, infor-metrics? What are we talking about? In: R.R.L. Egghe (ed.), *Informetrics*. Ontario, 89(90), pp. 31–43.

Caragliu, A., Del Bo, C. and Nijkamp, P. (2011). Smart cities in Europe. *Journal of Urban Technology*, 18(2), pp. 65–82. doi:10.1080/10630732.2011.601117.

Chen, C. (2014). *The CiteSpace Manual*, College of Computing and Informatics, pp. 1–84. doi:10.1007/s11192-015-1576-8.

Chen, C. and Song, M. (2019). Visualizing a field of research: A methodology of systematic scientometric reviews. *PLoS One*, 14(10), p. e0223994. doi:10.1371/journal.pone.0223994.

Chourabi, H., Nam, T., Walker, S., Gil-Garcia, J.R., Mellouli, S., Nahon, K. and Scholl, H.J. (2012). Understanding smart cities: An integrative framework. *System Science (HICSS), 2012 45th Hawaii International Conference*, IEEE, pp. 2289–2297. January 4–7, Hawaii.

Darko, A., Chan, A.P.C., Adabre, M.A., Edwards, D.J., Hosseini, M.R. and Ameyaw, E.E. (2020). Artificial intelligence in the AEC industry: Scientometric analysis and visualization of research activities. *Automation in Construction*, 112. doi:10.1016/j.autcon.2020.103081.

Freeman, L.C. (1977). A set of measures of centrality based on betweenness. *Sociometry*, 40(1), pp. 35–41. doi:10.2307/3033543.

Hood, W.W. and Wilson, C.S. (2001). The literature of bibliometrics, scientometrics, and informetrics. *Scientometrics*, 52(2), pp. 291–314. doi:10.1023/A:1017919924342.

Hosseini, M.R., Martek, I., Zavadskas, E.K., Aibinu, A.A., Arashpour, M. and Chileshe, N. (2018). Critical evaluation of off-site construction research: A Scientometric analysis. *Automation in Construction*, 87, pp. 235–247.

Jayasree, V. and Baby, M.D. (2019). Scientometrics: Tools, techniques, and software for analysis. *Indian Journal of Information Sources and Services*, 9(2), pp. 116–121.

Jin, R., Gao, S., Cheshmehzangi, A. and Aboagye-Nimo, E. (2018). A holistic review of off-site construction literature published between 2008 and 2018. *Journal of Cleaner Production*, pp. 202, 1202–1219. doi:10.1016/j.jclepro.2018.08.195.

Kathiravelu, P., Sharifi, L. and Veiga, L. (2015). Cassowary: Middleware platform for context-aware smart buildings with software-defined sensor networks. *Proceedings of the 2nd Workshop on Middleware for Context-Aware Applications in the IoT*, pp. 1–6. December 7–11, Vancouver.

Kleinberg, J. (2002). Bursty and hierarchical structure in streams. *Proceedings of the 8th ACM SIGKDD International Conference on Knowledge Discovery and Data Mining,* ACM SIGKDD, vol. 7, pp. 373–397. doi:10.1023/A:1024940629314.

Kramers, A., Höjer, M., Lövehagen, N. and Wangel, J. (2014). Smart sustainable cities – Exploring ICT solutions for reduced energy use in cities. *Environmental Modelling & Software,* 56, pp. 52–62. doi:10.1016/j.envsoft.2013.12.019.

Lazaroiu, G.C. and Roscia, M. (2012). Definition methodology for the smart cities model. *Energy,* 47(1), pp. 326–332.

Milenković, M., Rašić, M. and Vojković, G. (2017). Using public private partnership models in smart cities-proposal for Croatia. *40th International Convention on Information and Communication Technology, Electronics, and Microelectronics (MIPRO),* IEEE, May, pp. 1412–1417. May 22–26, Opatija.

Mooghali, A., Alijani, R., Karami, N. and Khasseh, A. (2011). Scientometric analysis of the scientometric literature. *International Journal of Information Science and Management,* 9, pp. 19–31.

Nam, T. and Pardo, T.A. (2011). Conceptualizing smart city with dimensions of technology, people, and institutions. *Proceedings of the 12th Annual International Digital Government Research Conference: Digital Government Innovation in Challenging Times,* pp. 282–291. Maryland.

Neirotti, P., De Marco, A., Cagliano, A.C., Mangano, G. and Scorrano, F. (2014). Current trends in smart city initiatives: Some stylised facts. *Cities,* 38, pp. 25–36. doi:10.1016/j.cities.2013.12.010.

Olawumi, T.O. and Chan, D.W.M. (2018). A scientometric review of global research on sustainability and sustainable development. *Journal of Cleaner Production,* 183, pp. 231–250. doi:10.1016/j.jclepro.2018.02.162.

Olawumi, T.O., Chan, D.W.M. and Wong, J.K.W. (2017). Evolution in the intellectual structure of BIM research: A bibliometric analysis. *Journal of Civil Engineering and Management,* 23(8), pp. 1060–1081. doi:10.3846/13923730.2017.1374301.

Oldewurtel, F., Parisio, A., Jones, C.N., Gyalistras, D., Gwerder, M., Stauch, V. and Morari, M. (2012). Use of model predictive control and weather forecasts for energy-efficient building climate control. *Energy and Buildings,* 45, pp. 15–27. doi:10.1016/j.enbuild.2011.09.022.

Oraee, M., Hosseini, M.R., Papadonikolaki, E., Palliyaguru, R. and Arashpour, M. (2017). Collaboration in BIM-based construction networks: A bibliometric-qualitative literature review. *International Journal of Project Management,* 35(7), pp. 1288–1301.

Peris-Ortiz, M., Bennett, D.R. and Yábar, D.P.B. (2017). Sustainable smart cities. In: M. Peris-Ortiz et al. (eds.) *Innovation, Technology, and Knowledge Management.* Springer International Publishing, Cham, 218 pages. doi: 10.1007/978-3-319-40895-8

Qiu, J., Zhao, R., Yang, S. and Dong, K. (2017). *Informetrics: Theory, Methods, and Applications,* Springer, Singapore.

Ramallo, J., Johnson, E. and Spencer Jr., B. (2002). "Smart" base isolation systems. *Journal of Engineering Mechanics,* 128(10), pp. 1088–1099.

Ramaprasad, A., Sánchez-Ortiz, A. and Syn, T. (2017, September). A unified definition of a smart city. In International Conference on Electronic Government. Springer, Cham. St. Petersburg, pp. 13–24. doi: 10.1007/978-3-319-64677-0_2

Rana, N.P., Luthra, S., Mangla, S.K., Islam, R., Roderick, S. and Dwivedi, Y.K. (2019). Barriers to the development of smart cities in Indian context. *Information Systems Frontiers,* 21(3), pp. 503–525.

Saka, A.B. and Chan, D.W.M. (2019a). A global taxonomic review and analysis of the development of BIM research between 2006 and 2017. *Construction Innovation*, 19(3), pp. 465–490. doi:10.1108/ci-12-2018-0097.

Saka, A.B. and Chan, D.W.M. (2019b). A scientometric review and metasynthesis of building information modelling (BIM) research in Africa. *Buildings*, 9(4), Art. 85, 21 pages. doi:10.3390/buildings9040085.

Schaffers, H., Komninos, N., Pallot, M., Trousse, B., Nilsson, M. and Oliveira, A. (2011, May). Smart cities and the future internet: Towards cooperation frameworks for open innovation. In: J. Domingue et al. (eds.) *The Future Internet Assembly*, Springer, Berlin, Heidelberg, vol. 6656, pp. 431–446. doi: 10.1007/978-3-642-20898-0_31

Siluo, Y. and Qingli, Y. (2017). Are scientometrics, informetrics, and Bibliometrics different? *ISSI 2017-16th International Conference on Scientometrics and Informetrics, Conference Proceedings*, ISSI, Wuhan, China, pp. 1507–1518.

Soyinka, O., Siu, K.W.M., Lawanson, T. and Adeniji, O. (2016). Assessing smart infrastructure for sustainable urban development in the Lagos metropolis. *Journal of Urban Management*, 5(2), pp. 52–64.

Spencer Jr., B. and Nagarajaiah, S. (2003). State of the art of structural control. *Journal of Structural Engineering*, 129(7), pp. 845–856.

Tague-Sutcliffe, J. (1992). An introduction to informetrics. *Information Processing and Management*, 28, pp. 1–3. doi:10.1016/0306-4573(92)90087-G.

Van Eck, N.J. and Waltman, L. (2014). Visualizing bibliometric networks. In: Y. Ding, R. Rousseau and D. Wolfram (eds.) *Measuring Scholarly Impact*, Springer, Cham, pp. 285–320. doi: 10.1007/978-3-319-10377-8_13

Van Eck, N.J. and Waltman, L. (2019). *VOSviewer manual*. Retrieved online from Leiden: Univeristeit Leiden. Available at: https://www.vosviewer.com/documentation/Manual_VOSviewer_1.6.8.pdf.

Yang, J.N. and Agrawal, A.K. (2002). Semi-active hybrid control systems for nonlinear buildings against near-field earthquakes. *Engineering Structures*, 24, pp. 271–280.

Zanella, A., Bui, N., Castellani, A., Vangelista, L. and Zorzi, M. (2014). Internet of things for smart cities. *IEEE Internet of Things Journal*, 1(1), pp. 22–32. doi:10.1109/jiot.2014.2306328.

Zhao, X. (2017). A scientometric review of global BIM research: Analysis and visualization. *Automation in Construction*, 80(April), pp. 37–47.doi:10.1016/j.autcon.2017.04.002.

12 Analysis of BIM-FM integration using a science mapping approach

Ecem Tezel and Heyecan Giritli

Introduction

Building Information Modelling (BIM) has been a fast growing phenomenon in the field of Architecture, Engineering, and Construction (AEC) Industry as a means of enhancing communication and co-ordination, reducing errors, omissions and re-work, improving productivity and quality, and decreasing cost and duration in all phases of the building life cycle (Suermann and Issa, 2009; Azhar, 2011; Bryde *et al.*, 2013; Love *et al.*, 2013; Zhou *et al.*, 2017). Essentially, BIM is an IT-enabled approach that involves applying and maintaining an integrated digital representation of all building information for different phases of the project life cycle in the form of a data repository (Gu and London, 2010). Facilities management (FM), however, refers to an integrated approach to operating, maintaining, improving, and adapting the buildings and infrastructure of an organisation in order to create an environment that strongly supports the primary objectives of that organisation (Barrett and Baldry, 2003). Involving a wide range of services from daily maintenance operation to long-term property management (Chotipanich, 2004; Atkin and Brooks, 2009), the FM profession encompasses multiple disciplines to ensure the functionality, comfort, safety, and efficiency of the built environment by integrating people, place, process, and technology (IFMA, 2019).

From the AEC point of view, FM is a set of services required during the operation and maintenance (O&M) phase of the buildings that includes maintenance and repair work, commissioning and replacement, energy management, emergency planning, space management, occupant comfort, and so on. It is generally acknowledged that buildings are designed or selected according to the needs and requirements of their occupants, but several FM practices are carried out throughout the building's occupancy period. Thus, the O&M phase can be said to be the most information-intensive phase in the building life cycle.

Undoubtedly, FM practices require complete, accurate, and available information, which is where BIM has emerged as a beneficial tool for the efficient management of information during O&M (Pärn *et al.*, 2017; Wong *et al.*, 2018; Gao and Pishdad-Bozorgi, 2019; Matarneh *et al.*, 2019a). BIM offers certain benefits for FM practitioners, including locating building components, accessing real-time facility data, visualisation and marketing, checking maintainability, creating and updating digital assets, managing space, carrying out planning and feasibility studies for non-capital construction, managing emergencies, monitoring and controlling energy, and training and developing personnel (Becerik-Gerber *et al.*, 2012). These benefits of BIM in FM practices derive mainly from its capacity to store geometric (i.e. shape, size, height) and non-geometric (i.e. cost, specification, manufacturer, installation date, warranty) information during the whole life cycle of the buildings. However, there are certain challenges also hindering the implementation of BIM in FM. The two fundamental challenges are technology oriented (i.e. inter-operability between different software systems) and problems and process oriented (i.e. poor definition of roles, responsibilities, and maintenance of digital information).

Until recently, the primary focus was on the use of BIM during design and construction phases. After 2010, however, the interest of global research has shifted from BIM-based design and construction to BIM-based FM (Gao and Pishdad-Bozorgi, 2019), and since that date, there has been an increasing number of studies on BIM-FM integration (Matarneh *et al.*, 2019a). Even though this is a relatively new topic, several attempts have been made to review these integration studies. However, the findings of these secondary research studies have shown certain limitations, such as specific and/or limited research focus (Wong *et al.*, 2018; Matarneh *et al.*, 2019a) and limited industry response (Dixit *et al.*, 2019). Moreover, in some of these studies, either only one scientific database is reviewed, namely Scopus or Web of Science (WoS) (Gao and Pishdad-Bozorgi, 2019; Matarneh *et al.*, 2019a), or a non-systematic approach is used (Yalcinkaya and Singh, 2014; Naghshbandi, 2016) during the research process.

It is clear that it has become necessary for the increasing number of BIM and FM studies to be researched overall. Furthermore, the given limitations of the previous studies have generated a need for a holistic approach. Therefore, the aim of this study was to fill this gap by offering a systematic review of the current empirical research on how the integration of BIM-FM has been studied and discussed in the literature. The purpose of the science mapping approach, applied in this chapter, was to create a holistic assessment of the existing knowledge of the implementation of BIM in FM practices by analysing contemporary trends and identifying research gaps for further studies. This study was expected to make two contributions of interest to readers. First, regardless of the subject, the application of science mapping, a popular secondary research method, has been demonstrated in this chapter. Second, the current state of BIM-FM integration has been discussed and future directions in research have been proposed.

Methodology

In this study, the science mapping approach, adapted from Jin *et al.* (2019), has been applied to: (1) analyse the influential keywords, articles, scholars, and journals in the domain of BIM and FM; (2) analyse the mainstream research topics in BIM-FM integration; (3) discuss the limitations or gaps of existing research in BIM-FM integration; and (4) propose directions for future research work

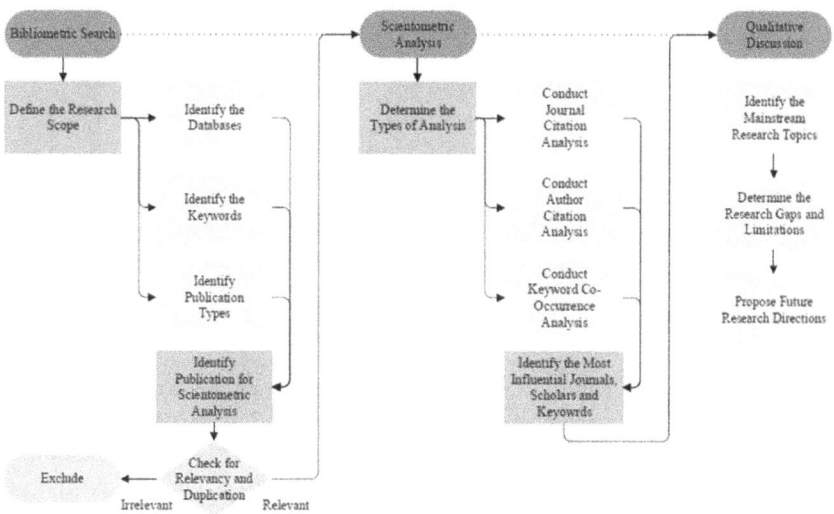

Figure 12.1 Research design and process.

(see Figure 12.1). The reason for selecting this procedure from other review approaches was that it is a more comprehensive review approach that can be incorporated into a qualitative, follow-up discussion. Different from previous review
studies in the BIM-FM domain, in this study, both WoS and Scopus, the two
main scientifically reputable databases, were scanned and VOSviewer was used to
visualise the results.

The science mapping process in this study included the following steps:

Step 1: Bibliometric search

A bibliometric search of the journal articles in the BIM and FM domains was
performed. Although there were different bibliometric sources available, such as
Scopus, WoS, EBSCOhost, or ProQuest, the data for this study were obtained
from two of them, Scopus and WoS. The reason for this selection was that, for
most scientific disciplines, these two databases are recommended often as being
the most important bibliographic databases because of their comprehensiveness
and scientific robustness.

Considering that using exact wording could result in missing some relevant
documents, different wording was used in the query string, including: "BIM" OR
"building information model" OR "building information modeling" OR "building information modelling" AND "FM" OR "facility management" OR "facilities management". Words in the title/abstract/keyword fields were searched to
find all related publications. This initial search identified 477 journal articles
published between 1998 and 2019 in English. Duplicates and irrelevant articles

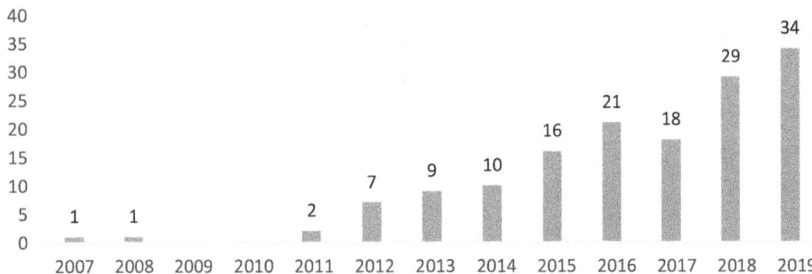

Figure 12.2 Distribution of the journal articles published in WoS and Scopus over years.

were determined by carefully reading their title/abstract/keywords and were removed during further screening. Finally, a total of 150 articles were selected for scientometric analysis. The detailed distribution of articles over years is shown in Figure 12.2.

Step 2: Scientometric analysis

Scientometric analysis of the selected articles was performed. Several visualisation software tools, such as CiteSpace, Gephi, Pajek, Science of Science Tool, BibExcel, and VOSviewer were available for rapid and effective scientometric analysis. For this study, VOSviewer software was used for visualisation. The main reasons for selecting VOSviewer were: it is a free software package developed specifically for displaying bibliometric maps, it uses Java programming language which operates on most hardware and operating systems, and it has a user-friendly interface with sophisticated calculations for advanced analysis (van Eck and Waltman, 2010, 2018). Börner *et al.* (2003) indicated that journals, documents, cited references, authors, and descriptive terms or keywords are the fundamental units of scientometric analysis. Thus, for this study, the following were performed: journal citation analysis to determine the most productive and influential journals of BIM and FM studies; author citation analysis to determine the influential scholars in the field; and keyword co-occurrence analysis to identify the mainstream research topics of BIM-FM integration.

Before proceeding to the analysis stage, the two different bibliometric data obtained from WoS, in tab-delimited format, and Scopus, in comma separated values (CSV) format, were combined for accurate analyses. Therefore, based on the required data for the selected analyses, the exported files from both databases were manipulated and combined into a single CSV file.

Step 3: Qualitative discussion

Qualitative discussion was the final step, following bibliometric search and scientometric analysis. As shown in Figure 12.1, three main objectives were to be achieved as a result of the in-depth, qualitative discussion: an in-depth evaluation of the mainstream research topics in BIM-FM integration; identification of current research gaps or limitations; and identification of future research directions.

Results and discussion

The initial results of the bibliometric analysis (Figure 12.2) showed the increasing trend of BIM and FM studies over the last decade, especially during the last five years. It is noted that the integration of BIM and FM fields is a relatively new topic, and more research in this domain is expected to be published in the near future (Matarneh *et al.*, 2019a).

Journal analysis

Journal articles are published after a detailed peer review process, and thus are accepted as reliable knowledge sources. The 150 articles analysed in this study were published in 57 distinguished journals between 2008 and 2019. A total of nine out of 57 journals met the minimum criteria of four articles and 20 citations. The journals are listed in Table 12.1, including the number of publications they have (No. of Pubs.), the number of citations (Total Citation), normalised citation (Norm. Citation), average citation (Avg. Citation), and average normalised citation (Avg. Norm. Cit.) values. The Norm. Citation value was provided by the VOSviewer software and the Avg. Norm. Cit. value was calculated by dividing the normalised citation of the related journal by the average number of citations of that journal (van Eck and Waltman, 2018).

Table 12.1 Quantitative summary of the most productive journals in BIM-FM research

Journal	No. of Pubs.	Total Citation	Norm. Citation	Avg. Citation	Avg. Norm. Cit.
Automation in Construction	26	946	45.89	36.38	1.76
Journal of Management in Engineering	4	99	6.63	24.75	1.66
Journal of Computing in Civil Engineering	7	146	7.95	20.86	1.13
Buildings	7	79	7.23	11.29	1.03
Built Environment Project and Asset Management	5	110	5.11	22.00	1.02
Advanced Engineering Informatics	8	181	7.24	22.63	0.90
Facilities	16	92	13.71	5.75	0.86
Journal of Performance of Constructed Facilities	5	32	2.34	6.40	0.47
Journal of Information Technology in Construction	6	52	2.22	8.67	0.37

Source: Original.

As shown in Table 12.1, *Automation in Construction* is the most productive journal with 26 articles, followed by *Facilities* (16), *Advanced Engineering Informatics* (8), *Journal of Computing in Civil Engineering* (7), and *Buildings* (7). Table 12.1 also shows that *Automation in Construction* is the most influential journal in BIM-FM integration studies based on the average normalised citations (Jin *et al.*, 2019), and followed by the *Journal of Management in Engineering* and *Journal of Computing in Civil Engineering.*

Scholar analysis

The author citation analysis was used to determine the most influential scholars in the BIM-FM community, according to the number of articles and citations they have. As shown in Figure 12.3, out of 405 authors in the community, a total of 24 authors met the minimum criteria of three articles and 20 citations. A detailed summary of the most influential authors is shown in Table 12.2.

Figure 12.3 demonstrates the high level of disconnectedness among 405 BIM-FM researchers. This disconnectedness might be the result of the various sub-domains of the FM field. However, collective knowledge generated by the mutual effort of the researchers in the same field could contribute to rapid development in the BIM-FM domain.

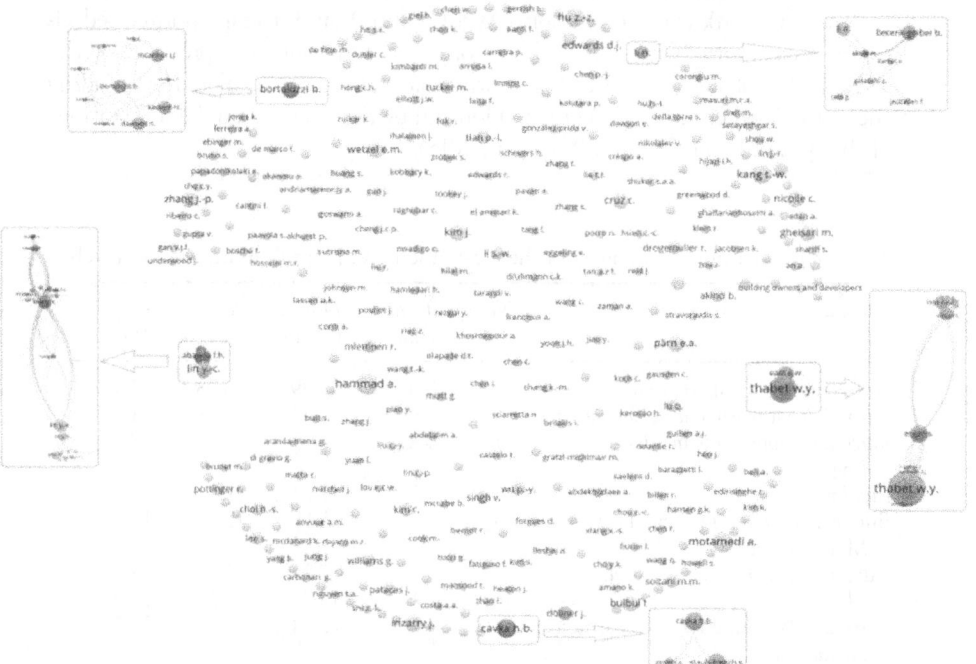

Figure 12.3 Visualisation of BIM-FM research community.

Note: node and font size represent the number of articles of each scholar in BIM-FM domain.

Table 12.2 Quantitative summary of the most influential scholars in BIM-FM community

Scholar	No. of Pubs.	Total Citation	Norm. Citation	Avg. Citation	Avg. Norm. Cit.
Issa, R.R.A.	3	77	5.73	25.67	1.91
Zhang, J.P.	3	45	4.23	15.00	1.41
Dawood, N.N.	3	124	4.20	41.33	1.40
Kassem, M.	3	124	4.20	41.33	1.40
Motamedi, A.	5	159	6.68	31.80	1.34
Gheisari, M.	3	86	3.50	28.67	1.17
Irizarry, J.	3	86	3.50	28.67	1.17
Edwards, D.J.	3	41	3.47	13.67	1.16
Hammad, A.	5	158	5.34	31.60	1.07
Hu, Z.Z.	4	45	4.23	11.25	1.06
Wetzel, E.M.	3	77	2.88	25.67	0.96
Nicolle, C.	3	235	2.88	78.33	0.96
Cavka, H.B.	4	54	3.38	13.50	0.84
Staun-French, S.	4	54	3.38	13.50	0.84
Yang, X.	3	26	1.92	8.67	0.64
Thabet, W.Y.	8	148	5.04	18.50	0.63
Ergan, S.	4	28	2.17	7.00	0.54
Bulbul, T.	3	59	1.30	19.67	0.43
Cruz, C.	3	179	1.28	59.67	0.43
Lucas, J.D.	6	71	2.16	11.83	0.36

Source: Original.

The scholars are listed in Table 12.2, including their total number of publications (No. of Pubs.), their number of citations (Total Citation), normalised citation (Norm. Citation), average citation (Avg. Citation), and average normalised citation (Avg. Norm. Cit.) values. Similar to the calculations made in the journal analysis, the Norm. Citation value was provided by the VOSviewer software and the Avg. Norm. Cit. value was calculated by dividing the normalised citation of each author by the average number of citations of that author (van Eck and Waltman, 2018).

As shown in Table 12.2, Issa was the most influential author based on average normalised citation (Avg. Norm. Cit.), followed by Zhang, Dawood, and Kassem. It could be concluded from this that the most influential studies in the BIM-FM domain were focused on FM requirements and maintainability problems (Liu and Issa, 2016), owner BIM competencies and information requirements (Giel and Issa, 2016; Mayo and Issa, 2016; Patacas *et al.*, 2016); use of open data standards, such as IFC and Cobie (Patacas *et al.*, 2015); challenges in adopting BIM in FM (Kassem *et al.*, 2015); efficient MEP management (Hu *et al.*, 2016, 2018); and data mining for information extraction (Peng *et al.*, 2017). The remaining studies covered a wide variety of issues, such as safety applications during the O&M phase (Pärn *et al.*, 2019), augmented reality (AR) integration (Williams *et al.*, 2015; Gheisari and Irizarry, 2016), or equipment localisation with radio frequency identification (RFID) tags (Motamedi *et al.*, 2013).

Keywords analysis

Keywords represent the main concepts researched for existing studies. Keyword co-occurrence analysis shows how frequently the use of a specific keyword occurred in literature and what the relationships are among those keywords. Additionally, the average normalised citation value of each keyword refers to how influential that keyword is in its research community. However, the keywords of studies are determined by their authors, and the unstandardised selection of keywords increases the number of keywords used in studies with similar content. One of the pre-processing applications suggested by Cobo *et al.* (2011) to obtain more accurate results from keyword co-occurrence analysis is to identify duplicates and misspelled items. This pre-processing helps researchers to identify the different spelling of a word for the same object or concept and to reduce the number of errors. In this study, initially, keywords with the same semantic meaning, such as "FM", "facility management", and "facilities management" or "BIM", "building information model", "building information modeling" and "building information modelling", were identified and combined as single terms of "FM" or "BIM". Then, the author keywords with a minimum of three occurrences were selected. Ultimately, 38 out of 418 keywords were mapped in Figure 12.4.

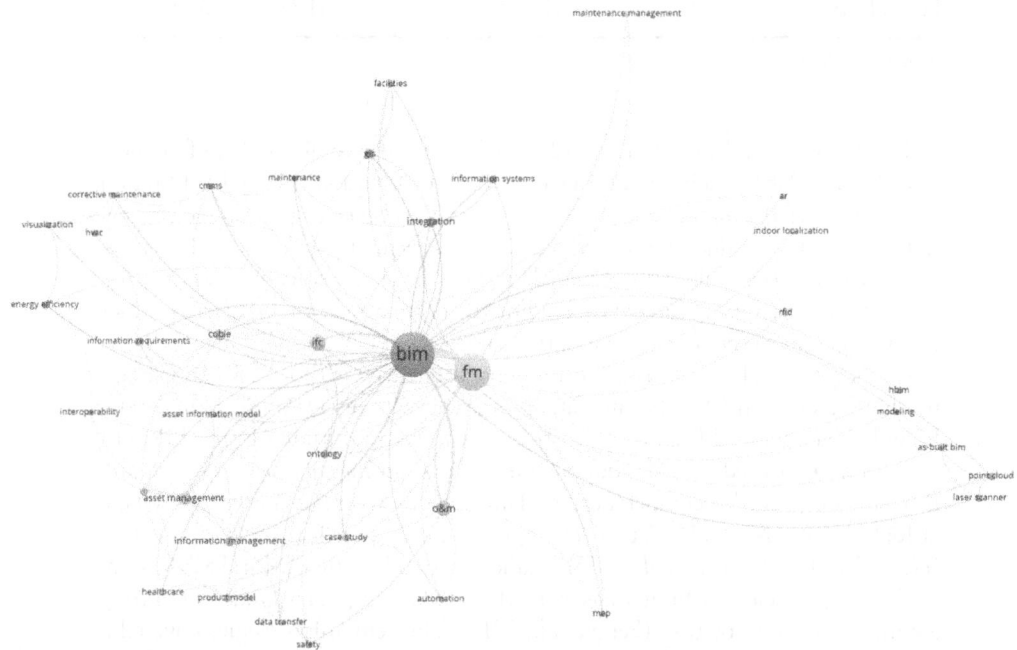

Figure 12.4 Visualisation of 38 out of 418 author keywords.

Note: That, node and font size represent the number of occurrences of each keyword and connection lines represent the inter-relatedness between keywords.

The keywords in Figure 12.4 are grouped under seven clusters according to their inter-relatedness. The focus of each cluster is explained briefly below:

- Cluster 1 is focused on the integration of various information systems, that is, BIM, Computerised Maintenance Management System, and Geographic Information System (GIS) for maintenance practices.
- Cluster 2 is focused on the data transfer/information exchange among different systems and facilities, that is, safety-related purposes or healthcare facilities.
- Cluster 3 is focused on corrective maintenance activities and covers the visualisation of heating, ventilation, and air conditioning systems and their energy performance.
- Cluster 4 is focused on maintenance management activities and related technologies, such as RFID tags and mobile AR applications.
- Cluster 5 is focused on the use of BIM and other technologies, such as laser scanners for ascertaining the as-built model of historical/heritage buildings.
- Cluster 6 is focused on inter-operability problems which is one of the main challenges to BIM-FM integration.
- Cluster 7 is focused on the case study method for real-life examples of automation in O&M practices.

The keywords are listed in Table 12.3, including their values for occurrence frequency (Occur. Freq.), average citation (Avg. Citation), average normalised

Table 12.3 Quantitative summary of the most influential keywords

Keyword	Occur. Freq.	Avg. Citation	Avg. Norm. Citation	Avg. Pub. Y.
AR	3	9.00	2.60	2018
Maintenance management	3	30.67	2.47	2016
Indoor localisation	3	6.67	2.19	2018
Point cloud	4	45.75	2.17	2016
Laser scanner	4	46.00	2.11	2015
HBIM	3	16.00	2.00	2018
Facilities	3	108.33	1.78	2015
Ontology	5	40.00	1.51	2014
Modelling	3	14.67	1.43	2017
O&M	15	16.20	1.29	2017
As-built BIM	3	39.33	1.25	2015
Automation	3	3.67	1.24	2018
Asset information model	3	3.33	1.21	2018
MEP	3	11.67	1.17	2018
Asset management	9	11.89	1.16	2018
FM	86	21.17	1.10	2016
COBie	8	20.38	1.08	2017
IFC	16	30.81	1.03	2015
GIS	5	29.60	1.00	2016
Information management	6	24.67	1.00	2016

Source: Original.

citation (Avg. Norm. Citation), and average publication year (Av. Pub. Y.). The main motivation for using the Av. Norm. Citation calculation is to prevent the dominance of older publications as a result of their time advantage in accumulating citations (van Eck and Waltman, 2018). However, the Av. Pub. Y. value denotes the average publication year of each keyword, which indicates how currently it was used. It is noted that keywords with Av. Norm. Citation over 1.00 were not listed in the table owing to limited space. The Av. Norm. Citation value of the BIM keyword, as one of the main keywords of this analysis, was 0.98, thus it was not listed in the table. It is shown in Table 12.3 that AR applications was the most influential research subject, based on its average normalised citation, followed by maintenance management, indoor localisation, point cloud, laser scanner, and Historic Building Information Modelling (HBIM).

Here, it can be concluded that AR applications in maintenance management, localization of building equipment, or the integrated use of BIM and laser scanner in heritage buildings maintenance are likely to be the higher impact studies in BIM-FM domain.

Qualitative discussion

The purpose of qualitative discussion is to determine the main research topics in a given field in order to identify the research gaps and propose future research directions. The clusters obtained from keyword co-occurrence analysis are an established starting point for identifying the mainstream research areas. However, as stated by Matarneh *et al.* (2019a), clustering might be insufficient for identifying relatively new topics. As a result, the main research topics in this case were categorised under four groups and the existing research gaps as well as potential future directions for each research topic are explained within each topic below:

- Topic 1: BIM implementation in FM. Together with the positive contribution that BIM brings to design and construction professionals, the main benefits of BIM are seen by the owners during the operational phase. In studies of this topic, the current status of BIM adoption in the industry is identified (Becerik-Gerber *et al.*, 2012) and the advantages and challenges of BIM application in FM are analysed (Kassem *et al.*, 2015; Pishdad-Bozorgi *et al.*, 2018). However, lack of case studies seemed to be the biggest gap in increasing awareness and adoption of BIM benefits within the industry. Therefore, further research effort should be focused on capturing and analysing more real-life, BIM-based FM examples.
- Topic 2: Information requirements, exchange, and management. It was clear that the operational phase of buildings takes significantly longer compared with design and construction. During this long period, efficient FM practices depend highly on complete and accurate information. Studies of this topic make an effort to understand BIM information requirements for the O&M phase (Becerik-Gerber *et al.*, 2012; East *et al.*, 2013; Korpela *et al.*, 2015; Mayo and Issa, 2016); develop information management frameworks (Alnaggar

and Pitt, 2019); and improve the information exchange process (Matarneh *et al.*, 2019b). The two main study gaps were the quality of handover information and inter-operability among software. Therefore, further research effort should be made to develop a holistic guideline for efficient information management. This guideline should identify the information needs of FM practices for different types of buildings, stakeholders responsible for information, and the data exchange process between different software tools. Another, further research effort should be made to capturing FM data and transferring knowledge between operational and design phases.

- Topic 3: FM operations in buildings. BIM helps facility managers during the decision-making process of various FM practices by providing real-time data flow for energy management and indoor conditioning, storing logs and manuals for maintenance operations, or locating all building equipment for repair and replacement work. Studies of this topic deal with corrective maintenance (Shalabi and Turkan, 2017), energy consumption monitoring (Oti *et al.*, 2016), emergency planning (Isikdag *et al.*, 2013), and visualisation of FM work orders (Williams *et al.*, 2015). Future research effort should be made to improve maintenance efficiency with BIM and AR/VR integration or to integrate the lessons learnt during the O&M phase into the pre-design phase of new projects.

- Topic 4: Existing/heritage buildings. Despite the increasing trend in BIM-based design and construction of new buildings, existing building stock as well as heritage buildings could benefit from the use of BIM together with various disruptive technologies (i.e. laser scanner and point cloud) in the O&M phase. Studies of this topic are focused on the automatic, as-built data generation of existing buildings via point clouds (Wang *et al.*, 2015) or other image-based technologies (Lu and Lee, 2017) and BIM-based performance assessment of historic buildings (Bruno *et al.*, 2018). The main challenge in this topic is the lack of fully automatic and accurate generation of as-built BIM. Therefore, future research should be focused on the use of various recognition techniques to obtain rich, semantic data and automatic data transfer from site investigations to a central BIM model.

Conclusions

An increasing amount of research in a particular domain generates a need for a synthesising study to identify, evaluate, and interpret the knowledge that has been gained at any given time. Commonly known as systematic literature review, different methods and techniques are used in secondary research to examine the existing studies. In this chapter, the use of science mapping, a transparent, repeatable, and objective method, has been explained by using an example of a review of BIM-FM integration, which is one of the emerging topics in the built environment. The purpose of the science mapping approach applied in this chapter was to determine the most influential keywords, articles, scholars, and journals, to identify the main areas of research, to identify research gaps, and to propose topics for future research in the BIM-FM domain. First, a total of 150 journal articles published between 2008 and 2019 in WoS and Scopus databases were identified. Then, those articles were analysed and visualised with the help of VOSviewer software. Finally, existing research gaps as well as associated research directions were outlined.

The initial findings of this chapter showed that researchers started to concentrate on BIM-FM integration in 2008 and the number of studies has increased over the years. The results of the scientometric analyses showed that *Automation in Construction* was the most productive and influential journal, and Issa, Zhang, Dawood, and Kassem were the most influential authors in the field. Being a relatively new research domain, both the number of keywords and their frequency were limited. However, it appeared that the focus of research has been shifted from measuring awareness of the benefit of BIM and strategies to adopt BIM to: (1) BIM and AR integration for FM, (2) BIM-based maintenance management, and (3) use of point cloud and laser scanner to study existing buildings in general and historical buildings in particular.

Following the advances in data science and computer technology, different visualisation tools have been developed to streamline the systematic literature review process. These tools, which effectively display the relationships between different objects and concepts, help researchers to identify the research gaps that will guide further studies quickly and easily. The methodology and findings used in this chapter were expected to (1) provide a brief guideline for any researcher who is willing to conduct a science mapping study in any scientific field, and (2) help BIM-FM researchers to develop the focus of their research and strategy in terms of implementing BIM in FM practices.

References

Alnaggar, A. and Pitt, M. (2019). Lifecycle exchange for asset data (LEAD): A proposed process model for managing asset dataflow between building stakeholders using BIM open standards. *Journal of Facilities Management*, 17(5), pp. 385–411.

Atkin, B. and Brooks, A. (2009). *Total Facilities Management*, 3rd edition, Blackwell Publishing, Chichester.

Azhar, S. (2011). Building information modeling (BIM): Trends, benefits, risks and challenges for the AEC Industry. *Leadership and Management in Engineering*, 11(3), pp. 241–252.

Barrett, P. and Baldry, D. (2003). *Facilities Management: Towards Best Practice*, 2nd edition, Blackwell Publishing, Chichester.

Becerik-Gerber, B., Jazizadeh, F., Li, N. and Calis, G. (2012). Application areas and data requirements for BIM-enabled facilities management. *Journal of Construction Engineering and Management*, 138(3), pp. 431–442.

Börner, K., Chen, C. and Boyack, K. (2003). Visualizing knowledge domains. *Annual Review of Information Science and Technology*, 37, pp. 179–255.

Bruno, S., De Fino, M. and Fatiguso, F. (2018). Historic building information modelling: Performance assessment for diagnosis-aided information modelling and management. *Automation in Construction*, 86, pp. 256–276.

Bryde, D., Broquetas, M. and Volm, J. (2013). The project benefits of building information modelling (BIM). *International Journal of Project Management*, 31, pp. 971–980.

Chotipanich, S. (2004). Positioning facility management. *Facilities*, 22, pp. 364–372.

Cobo, M.J., Lopez-Herrera, A.G., Herrera-Viedma, E. and Herrera, F. (2011). Science mapping software tools: Review, analysis and cooperative study among tools. *Journal of American Society for Information Science and Technology*, 62(7), pp. 1382–1402.

Dixit, M.K., Venkatraj, V., Ostadalimakhmalbaf, M., Pariafsai, F. and Lavy, S. (2019). Integration of facility management and building information modeling (BIM): A review of key issues and challenges. *Facilities*, 37(7/8), pp. 455–483.

East, E.W., Nisbet, N. and Liebich, T. (2013). Facility management handover model view. *Journal of Computing in Civil Engineering*, 27(1), pp. 61–67.

Gao, X. and Pishdad-Bozorgi, P. (2019). BIM-enabled facilities operation and maintenance: A review. *Advanced Engineering Informatics*, 39, pp. 227–247.

Gheisari, M. and Irizarry, J. (2016). Investigating human and technological requirements for successful implementation of a BIM-based mobile augmented reality environment in facility management practices. *Facilities*, 34(1/2), pp. 69–84.

Giel, B. and Issa, R.R.A. (2016). Framework for evaluating the BIM competencies of facility owners. *Journal of Management in Engineering*, 32(1), p. 04015024.

Gu, N. and London, K. (2010). Understanding and facilitating BIM adoption in the AEC Industry. *Automation in Construction*, 19(8), pp. 988–999.

Hu, Z.-Z., Tian, P.-L., Li, S.-W. and Zhang, J.P. (2018). BIM-based integrated delivery technologies for intelligent MEP management in the operation and maintenance phase. *Advances in Engineering Software*, 115, pp. 1–16.

Hu, Z.-Z., Zhang, J.-P., Yu, F.-Q., Tian, P.-L. and Xiang, X.-S. (2016). Construction and facility management of large MEP projects using a multi-scale building information model. *Advances in Engineering Software*, 100, pp. 215–230.

International Facility Management Association (IFMA). (2019). *What is facility management*. Available at: https://www.ifma.org/what-is-facility-management [Accessed on: 19 November 2019].

Isikdag, U., Zlatanova, S. and Underwood, J. (2013). A BIM-oriented model for supporting indoor navigation requirements. *Computers Environment and Urban Systems*, 41, pp. 112–123.

Jin, R., Zou, P.X.W., Piroozfar, P., Wood, H., Yang, Y., Yan, L. and Hani Y. (2019). A science mapping approach based review of construction safety research. *Safety Science*, 113, pp. 285–297.

Kassem, M., Kelly, G., Dawood, N., Serginson, M. and Lockley, S. (2015). BIM in facilities management applications: A case study of a large university complex. *Built Environment Project and Asset Management*, 5(3), pp. 261–277.

Korpela, J., Miettinen, R., Salmikivi, T. and Ihalainen, J. (2015). The challenges and potentials of utilizing building information modelling in facility management: The case of the center for properties and facilities of the University of Helsinki. *Construction Management and Economics*, 33(1), pp. 3–17.

Liu, R. and Issa, R.R.A. (2016). Survey: Common knowledge in BIM for facility maintenance. *Journal of Performance of Constructed Facilities*, 30(3), p. 04015033.

Love, P., Simpson, I., Hill, A. and Standing, C. (2013). From justification to evaluation: Building information modeling for asset owners. *Automation in Construction*, 35, pp. 208–216.

Lu, Q. and Lee, S. (2017). Image-based technologies for constructing as-is building information models for existing buildings. *Journal of Computing in Civil Engineering*, 31(4), p. 04017005.

Matarneh, S.T., Danso-Amoako, M., Al-Bizri, S., Gaterell, M. and Matarneh, R. (2019a). Building information modeling for facilities management: A literature review and future research directions. *Journal of Building Engineering*, 24, p. 100755.

Matarneh, S.T., Danso-Amoako, M., Al-Bizri, S., Gaterell, M. and Matarneh, R. (2019b). BIM-based facilities information: Streamlining the information exchange process. *Journal of Engineering Design and Technology*, 17(6), pp. 1304–1322.

Mayo, G. and Issa, R.R.A. (2016). Nongeometric building information needs assessment for facilities management. *Journal of Management in Engineering*, 32(3), p. 04015054.

Motamedi, A., Soltani, M.M. and Hammad, A. (2013). Localization of RFID-equipped assets during the operational phase of facilities. *Advanced Engineering Informatics*, 27(4), pp. 566–579.

Naghshbandi, S.N. (2016). BIM for facility management: Challenges and research gaps. *Civil Engineering Journal*, 2(12), pp. 679–684.

Oti, A.H., Kurul, E., Cheung, F. and Tah, J.H.M. (2016). A framework for the utilization of building management system data in building information models for building design and operation. *Automation in Construction*, 72, pp. 195–210.

Pärn, E.A., Edwards, D.J., Riaz, Z., Mehmood, F. and Lai, J. (2019). Engineering-out hazards: Digitising the management working safety in confined spaces. *Facilities*, 37(3/4), pp. 196–215.

Pärn, E.A., Edwards, D.J. and Sing, M.C.P. (2017). The building information modeling trajectory in facilities management: A review. *Automation in Construction*, 75, pp. 45–55.

Patacas, J., Dawood, N., Greenwood, D. and Kassem, M. (2016). Supporting building owners and facility managers in the validation and visualisation of asset information models (AIM) through open standards and open technologies. *Journal of Information Technology in Construction*, 21, pp. 434–455.

Patacas, J., Dawood, N., Vukovic, V. and Kassem, M. (2015). BIM for facilities management: Evaluating BIM standards in asset register creation and service life planning. *Journal of Information Technology in Construction*, 20, pp. 313–331.

Peng, Y., Lin, J.-R., Zhang, J.-P. and Hu, Z.-Z. (2017). A hybrid data mining approach on BIM-based building operation and maintenance. *Building and Environment*, 126, pp. 483–495.

Pishdad-Bozorgi, P., Gao, X., Eastman, C. and Self, A.P. (2018). Planning and developing facility management-enabled building information model (FM-enabled BIM). *Automation in Construction*, 87, pp. 22–38.

Shalabi, F. and Turkan, Y. (2017). IFC BIM-based facility management approach to optimize data collection for corrective maintenance. *Journal of Performance of Constructed Facilities*, 31(1), pp. 04016081.

Suermann, P. and Issa, R.R.A. (2009). Evaluating industry perceptions of building information modeling (BIM) impact on construction. *Journal of Information Technology in Construction (ITcon)*, 14, pp. 574–594.

van Eck, N.J. and Waltman, L. (2010). Software survey: VOSviewer, a computer program for bibliometric mapping. *Scientometrics*, 84, pp. 523–538.

van Eck, N.J. and Waltman, L. (2018). *VOSviewer Manual*, Univeristeit Leiden, Leiden, pp. 1–49.

Wang, C., Cho, Y.K. and Kim, C. (2015). Automatic BIM component extraction from point clouds of existing buildings for sustainability applications. *Automation in Construction*, 56, pp. 1–13.

Williams, G., Gheisari, M., Chen, P.J. and Irizarry, J. (2015). BIM2MAR: An efficient BIM translation to mobile augmented reality applications. *Journal of Management in Engineering*, 31(1), p. A4014009.

Wong, J.K.W., Ge, J. and He, S.X. (2018). Digitisation in facilities management: A literature review and future research directions. *Automation in Construction*, 92, pp. 312–326.

Yalcinkaya, M. and Singh, V. (2014). Building information modeling (BIM) for facilities management-literature review and future needs. In: S. Fukuda et al. (eds.), *IFIP International Conference on Product Lifecycle Management*, July 7–9, Springer, Yokohama, Japan, pp. 1–10.

Zhou, Y., Ding, L., Rao, Y., Luo, H., Medjdoub, B. and Zhong, H. (2017). Formulating project-level building information modeling evaluation framework from the perspectives of organizations: A review. *Automation in Construction*, 81, pp. 44–55.

13 Trends in recycled concrete research

A bibliometric analysis

Olalekan Shamsideen Oshodi
and Bankole Osita Awuzie

Introduction

The potential for improved levels of material circularity in the construction industry to contribute towards the attainment of a circular economy has been reported (Adams *et al.*, 2017; Arora *et al.*, 2019; Geldermans, 2016; Wang *et al.*, 2017). The construction industry has a reputation for negating sustainable development (SD) tenets and by extension, the circular economy, owing to the plethora of anthropogenic activities associated with the industry (Awuzie and Monyane, 2020). The industry's commitment towards ameliorating the negative impacts of its activities on the attainment of society's SD ethos has culminated in the evolution of taxonomies such as green construction, sustainable construction, and, most recently, circular construction. To enable a shift towards more sustainable production practices, concerted efforts have been channelled towards construction processes and transformation of materials (Ghaffar *et al.*, 2020). The use of alternative materials is increasingly becoming the norm in the industry. Concrete remains a dominant material used to produce the built environment and associated components. As a result of its predominance, most construction waste accrues from concrete. Therefore, identifying and adopting new approaches to the re-use of concrete in the construction industry will contribute significantly towards fostering circular construction values (Ogunmakinde *et al.*, 2017, Heeren and Hellweg, 2019). Several projects have been initiated to buttress the imperative nature of concrete waste management. Some of these initiatives include the Advanced Technologies for the Production of Cement and Clean Aggregates from Construction and Demolition Waste Project, and the Innovative Strategies for High-Grade Material Recovery from Construction and Demolition Waste Project (Ghaffar *et al.*, 2020).

Also, an increasing number of research projects focused efforts to manage the waste of concrete and concrete aggregates has been observed (Malešev *et al.*, 2010; McNeil and Kang, 2013). Given the rapid increase in publications focused on this niche area, a review of relevant publications within this knowledge domain has become imperative as this is currently lacking. It is expected that such a review of published research will engender an in-depth knowledge of the evolutionary trends associated with this knowledge domain. Furthermore, such an assessment will make it possible to identify knowledge gaps and future research directions. This was the rationale for carrying out the bibliometric analysis reported in this chapter.

The choice of the bibliometric analysis method was predicated on its contributions towards enabling a quantitative assessment of the communication of scientific output concerning a field of study (Zupic and Čater, 2015). Subsequent sections of the chapter will serve to illustrate, in a step-by-step manner, the processes adopted in carrying out a bibliometric analysis of the research about recycled concrete over a period of 35 years between the year of the first publication in 1983 and 2018.

The bibliometric analysis process

According to De Bellis (2009), bibliometrics originated from the analysis of quantitative patterns relating to a cluster of scientific documents detailing the works of scientists within a given domain. Martín-Martín *et al.* (2016, p. 3) defined bibliometrics as the:

> …discipline responsible for measuring communication and, in enlarged form, as a specialty responsible for quantitatively studying the production, distribution, dissemination and consumption of information conveyed in any type of document and any intellectual field but with special attention to scientific information.

An apparent lack of consensus concerning a widely accepted protocol for bibliometric analysis has been observed. Therefore, in conducting the bibliometric analysis of research about recycled concrete, the protocol developed by Zupic and Čater (2015) was used in this chapter. Zupic and Čater (2015) proposed a five-stage approach to conducting bibliometric analysis that comprises the following steps: the conceptualisation of the research and choice of method, elicitation of bibliographic information (data), analysis of collected data, visualisation, and presentation of the results. These five steps are explained below.

Step 1: Conceptualisation of research and choice of method

In the first phase of the study, the aim and objectives were clearly stipulated and the choice of an appropriate method for achieving these objectives was decided upon. The rationale behind the conceptualisation of the study has been provided in the previous section. The choice of an appropriate method was made from a set of alternatives, namely a systematic review of literature, meta-analysis, and bibliometric analysis. These alternatives constitute three main types of methods deployed for the synthesis of previous research (Gradeci *et al.*, 2017; Schmidt, 2008; Zupic and Čater, 2015). Apart from the usefulness of bibliometric analysis in mapping the structure and evolution of knowledge relating to research about recycled concrete, the choice of this method was predicated on its capacity to engender an objective analysis of previous research and reproducibility (Allison *et al.*, 2018).

Step 2: Gathering of bibliometric information

The second step entailed the collection of bibliographic information from a plethora of searches conducted on the selected database. Scholars seem to agree on the reliability of bibliometric data obtained from SCOPUS and Web of Science (WoS), when compared with the data from Google Scholar (GS), especially when the study being conducted is situated within the domain of natural and formal sciences (Mingers and Leydesdorff, 2015). Notwithstanding the wide coverage of GS as a database, scholars have continued to downplay its capacity to provide accurate and consistent data for bibliometric analysis. Martín-Martín *et al.* (2016) identified incorrect attribution of documents to authors, wrong indexing of scientific communication, inability to merge all versions of the same document into a single record, thereby resulting in multiple variants of the same document, as some of the mistakes which undermine the capacity of GS to serve as a veritable database for sourcing bibliographic information. Also, in comparing the performance of SCOPUS and WoS databases for sourcing bibliographic information, Falagas *et al.* (2008) maintained that SCOPUS offers a more comprehensive platform for accessing indexed journal articles. For purposes of credibility, journal article publications should be given priority when searching for relevant scientific communications for bibliometric analysis (De Oliveira *et al.*, 2019). The rationale behind this suggestion stems from the peer-reviewed reputation of such publications.

For this study, the SCOPUS database was selected, considering its reputation as a more comprehensive source of indexed journal articles (Falagas *et al.*, 2008, Mongeon and Paul-Hus, 2016). Mongeon and Paul-Hus (2016) asserted that SCOPUS is more inclined to support Natural Sciences, Engineering, and Biomedical research to the detriment of research from the Social Sciences, and Arts, and Humanities owing to the introduction of certain biases. Also, Mongeon and Paul-Hus (2016) mentioned that this database favours English-language journals over journals published in other languages.

After the adoption of a veritable database (SCOPUS), decisions were taken concerning appropriate keywords to use for optimal search results. The choice of keywords is critical as they enable the identification of relevant journal articles from the search of the database. For this study, the following keyword combinations were used for the SCOPUS search: (i) "recycled concrete"; (ii) "recycle*" AND "concrete". The search results were filtered to exclude the following: (i) duplicates and (ii) papers published in other languages apart from English. Also, the search results were limited to journal papers. See Table 13.1 for the inclusion and exclusion criteria.

In total, 1,018 relevant publications were identified at the end of the search process for research about recycled concrete. The bibliographic data relating to these publications were subsequently downloaded and stored for further analysis.

Table 13.1 Inclusion and exclusion criteria for database search

	Inclusion criteria	*Exclusion criteria*
Database	SCOPUS	Other databases
Publication period	Up to 2018	Articles published after 2018
Document type	Articles and reviews	Conference papers, book chapters, notes, letters, editorials
Source type	Journal	Books, websites, conference proceedings, trade publications, Doctoral thesis, Masters dissertations
Subject area	Construction and Building Technology Engineering, Civil Engineering Materials Science, Composites, Engineering Environmental Sciences Green and Sustainable Science and Technology	Others
Language	English	–

Source: Original.

Steps 3 and 4: Data analysis and visualisation

The measurement of scientific output and/or communication through bibliometric analysis is focused on various aspects of productivity as common units of analysis. These aspects may include any or a combination of the following: author productivity, research team productivity, journal productivity, knowledge domain state of the art as well as the geographical distribution. Different measurement indicators have been highlighted in several bibliometric studies. For instance, data relating to research areas can serve as an indication of the most searched areas in a database, whereas the titles of articles can be deployed to establish the most cited articles. Furthermore, the number of articles concerning a phenomenon, published in a journal, serves as an indication of the journal's level of productivity, whereas the identification of authors of relevant articles highlights the most cited (productive) authors in the knowledge domain being studied. For this study, certain units of analysis were adopted, including the number of publications per year, author distribution, and collaboration patterns. These units have been discussed in the appropriate sections.

VOSviewer was used in carrying out further analysis of the study's results. Three main tools are available for mapping and visualising bibliometric data, namely CiteSpace, HistCite, and VOSviewer (Pan *et al.*, 2018). VOSviewer enjoys considerable levels of prominence for analysis and visualisation of bibliometric data when compared with CiteSpace and HistCite, hence its prevalent usage (Pan *et al.*, 2018, van Eck and Waltman, 2013). For guidance on the functionality and usability of VOSviewer, see van Eck and Waltman (2013).

Step 5: Presentation of results

In this section, the results from the bibliometric analysis of research about recycled concrete are presented.

Number of publications per year

The temporal distribution of publications containing research about recycled concrete is summarised and presented in Figure 13.1. Based on the information available in the summary, the first paper on recycled concrete was published in 1983. Figure 13.1 also shows that the number of publications on this subject matter grew from one per year in 1983 to 226 per year by the end of 2018. Figure 13.1 provides insights into the evolution of research interest in recycled concrete, which can be categorised according to three time intervals: (i) initial stage from 1983 to 2000 (less than two publications per year); (ii) slow growth stage from 2001 to 2006 (between three and nine publications per year); and (iii) rapid growth stage (between 11 and 226 publications per year. Van Nunen *et al.* (2018) suggested that the number of publications serves as an indicator of the emergence of a topical subject area. A perusal of the content of Figure 13.1 seems to support this assertion, as the cumulative number of publications grew from one in 1983 to 1,018 at the end of 2018.

Distribution of authors and collaboration patterns

The 1,018 publications identified as relating to recycled concrete were written by 2,114 different authors. However, whereas a large proportion of authors published

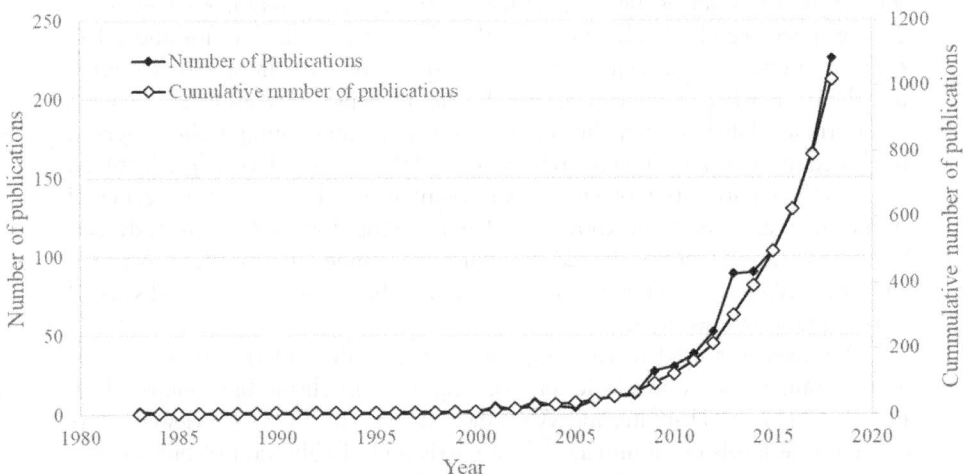

Figure 13.1 Number of publications about recycled concrete and cumulative number of publications about recycled concrete per year.

only one paper on recycled concrete (73.32%, n = 1,550/2,114), 119 authors (5.63%, n = 119/2,114) had published at least three papers. Furthermore, 20 authors (0.95%) had published at least ten or more journal articles. These findings were consistent with observations presented in previous studies (Goyal, 2017; van Nunen *et al.*, 2018), which showed that a small group of authors published a significant proportion of journal articles on a topic.

Out of the 2,114 authors, the top ten most productive authors, in terms of the numbers of journal papers published on recycled concrete, were identified (see Table 13.2).

Based on the information in Table 13.2, De Brito was the most productive author on the topic of recycled concrete with 42 publications. The second- and third-ranked authors, Arulrajah and Poon, had published 26 and 23 journal papers respectively. However, the top three most-cited authors were De Brito, Evangelista, and Poon with 1,875, 1,070, and 842 citations respectively. This unit of analysis is significant because it enables the researcher to develop an appreciation of the leading (seminal) authors within the knowledge domain.

Collaboration network

The collaboration network serves as another unit of analysis deployed in this study and within the context of bibliometric analysis. This makes it possible to determine the patterns of collaboration that exist on three different planes, namely authorship, institutions, and countries. This information is important for determining the patterns of knowledge flow between the main researchers and their collaborators who are working on the phenomenon, and for appraising the geographical coverage and institutions represented in such collaborative networks.

Table 13.2 Top ten of the most productive authors

No.	Author name	Country	Number of publications	Number of citations	Average citations per publications
1	De Brito, J.	Portugal	42	1,875	44.64
2	Arulrajah, A.	Australia	26	695	26.73
3	Poon, C.S.	Hong Kong	23	842	36.61
4	Gonzalez-Fonteboa, B.	Spain	20	436	21.80
5	Singh, B.	India	20	122	6.10
6	Martinez-Abella, F.	Spain	18	394	21.89
7	Perez, I.	Spain	17	238	14.00
8	Evangelista, L.	Portugal	15	1,070	71.33
9	Horpibulsuk, S.	Thailand	14	304	21.71
10	Singh, S.P.	India	13	110	8.46

Source: Original.

Authorship

The collaboration pattern (i.e. authorship) that existed among authors publishing journal papers on recycled concrete was analysed using VOSviewer. Authors included in this network published at least five papers on recycled concrete. Authors who were not connected with other authors were not included in the network. The pattern of collaboration among authors is presented in Figure 13.2. The size of the nodes indicates the number of published journal articles. The lines between the nodes represent collaboration between authors and the colours represent the collaboration cluster. Based on Figure 13.2, it was evident that De Brito and Gonzalez-Fonteboa were the main researchers in the network. Other researchers were linked to these main researchers.

In Figure 13.2, five main clusters (purple, blue, red, yellow, and green) were evident. The authors in the blue and purple cluster were affiliated to institutions located in Portugal, apart from Dhir who was affiliated to the University of Dundee in the UK. The authors in the red and green clusters were affiliated to Spanish institutions. The yellow cluster was made up of authors affiliation to institutions located in Italy, except for Letelier who was affiliated to the Universidad de la Frontera, Temuco, in Chile. Although the implication from Figure 13.2 was that authors were collaborating actively on research about recycled concrete, it was evident that most of the authors were affiliated to institutions located in

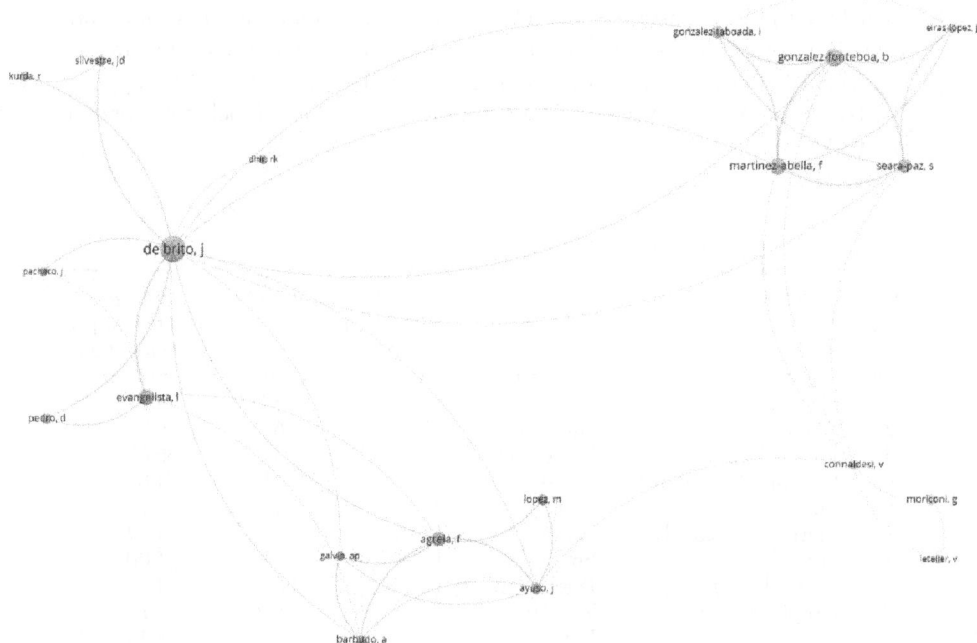

Figure 13.2 Collaboration network among researchers of recycled concrete.

Europe. This finding highlighted the need for more international collaborations and inter-disciplinary research focused on recycled concrete, which is a gap that can be explored by subsequent research projects.

Institution

The authors contributing to knowledge about recycled concrete are affiliated to 835 institutions (one author can be affiliated to more than one institution or several authors working at various institutions can contribute to the same manuscript). The use of fractional counting was recommended for analysis of the co-authorship network (Perianes-Rodriguez *et al.*, 2016). This approach was used in this study. Sivertsen *et al.* (2019) acknowledged the existence of two main approaches to counting article authorship data, namely full counting and fractional counting. The former accords one full credit to an author, the latter provides for the equal sharing of the credit among the contributing authors (Sivertsen *et al.*, 2019). By implication, the adoption of the fractional approach allowed for the sharing of publication credit between the number of authors in such a manner that each collaborating institution received a fraction instead of a full credit.

The top ten most active institutions in terms of the number of published studies about recycled concrete are presented in Table 13.3. The most productive institution was in Spain. This university published 43 journal papers. Interestingly, all of the top ten institutions were universities. This was indicative of the role of universities in shaping the evolving discourse about recycled concrete.

Countries

To gain an understanding of the geographical coverage of publications related to research about recycled concrete, the co-authorship network was analysed

Table 13.3 Top ten of the most active institutions in research about recycled concrete

No.	Institution	Country	Number of publications
1	Universidade da Coruña	Spain	43
2	Swinburne University of Technology	Australia	27
3	Hong Kong Polytechnic University	Hong Kong, China	27
4	University of Lisbon	Portugal	27
5	Polytechnic University of Catalonia	Spain	18
6	Universidad de Córdoba	Spain	15
7	Suranaree University of Technology	Thailand	14
8	Indian Institute of Technology, Roorkee	India	14
9	Tongji University	China	14
10	Nanyang Technological University	Singapore	14

Source: Original.

using VOSviewer. This analysis revealed that authors from 70 countries had contributed to this research topic within the period under review. The countries included in this data map, shown in Figure 13.3, published at least 15 papers on recycled concrete (18 countries met this threshold). Countries that are not connected to other countries were not included in this network. The size of the node serves as an indication of the number of papers published. The thickness of the link represents the number of collaborations among countries. The colour indicates collaboration clusters. Based on information available in Figure 13.3, there were four clusters (blue, yellow, red, and green). The green cluster consisted mainly of European Nations. The Red cluster consisted of countries located in Asia. The clusters tended to be centred around the most productive countries in research about recycled concrete. Similar to the findings in previous studies, collaborators tended to be located within the same geographical location (van Nunen *et al.*, 2018).

Journals that publish research about recycled concrete

In total, the 1,018 publications identified were published in 208 journals. The high number of outlets suggests that research about recycled concrete is multi-disciplinary in nature. Of the 208 journals, 106 journals published only one article culminating in 10.41% of the 1,018 journal papers reviewed, 40 journals

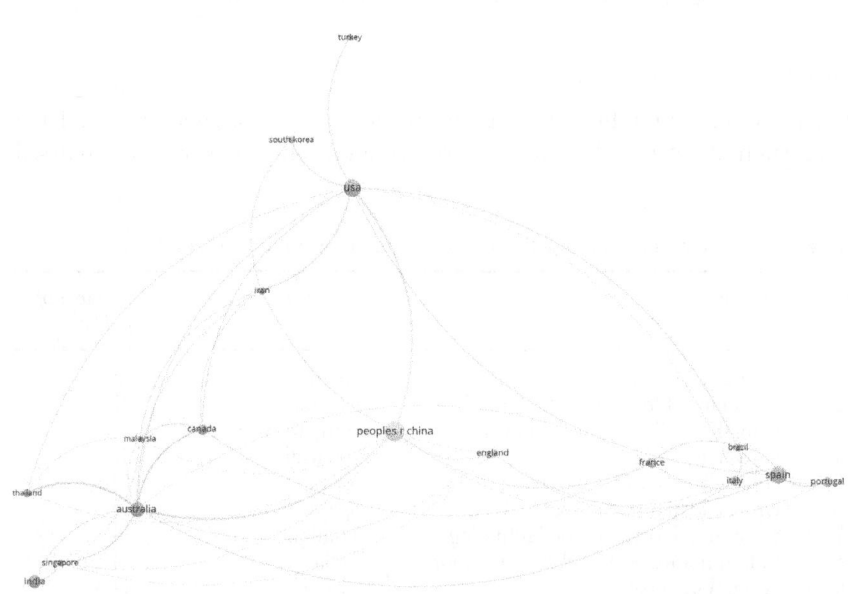

Figure 13.3 Collaboration network among countries involved in research about recycled concrete.

Table 13.4 Top ten of the most active journals publishing research about recycled
concrete

No.	Journal title	NP	NC	IF	SC
1t	Construction and Building Materials	257	6,789	4.046	Construction and Building Technology Engineering, Civil Materials Science, Multi-Disciplinary
2	Journal of Cleaner Production	56	1,903	6.395	Engineering, Environmental Environmental Sciences Green and Sustainable Science and Technology
3	Journal of Materials in Civil Engineering	51	1,198	1.984	Construction and Building Technology Engineering, Civil Materials Science, Multi-Disciplinary
4	Cement and Concrete Composites	38	1,599	5.172	Construction and Building Technology Materials Science, Composites
5	Materials and Structures	30	647	2.023	Construction and Building Technology Materials Science, Multi-Disciplinary
6	Transportation Research Record	26	135	0.748	Engineering, Civil Transportation Science and Technology
7	Magazine of Concrete Research	24	307	2.026	Construction and Building Technology Materials Science, Multi-Disciplinary
8	Waste Management	21	808	5.431	Engineering, Environmental Environmental Sciences
9	Cement and Concrete Research	20	1,599	5.613	Construction and Building Technology Materials Science, Multi-Disciplinary
10	ACI Materials Journal	20	365	1.453	Construction and Building Technology Materials Science, Multi-Disciplinary

Source: Original.
Note: NP = number of publications; NC = number of citations; IF = impact factor (2018 Journal
Citation Reports®), and SC = subject category of journal (2018 Journal Citation Reports®).

published only two articles on recycled concrete (7.86% of 1,018 journal papers),
and 16 journals published ten or more articles on this topic (61.20% of all papers).

Table 13.4 gives information about the top ten, peer-reviewed journals actively publishing research about recycled concrete. These top ten journals
published 543 articles, hence accounting for 53.34% of the global outputs.
Amongst these journals, *Construction and Building Materials* published the
highest number of articles on recycled concrete (25.25%; n = 257/1,018). The
other key journals in the field were *Journal of Cleaner Production* and *Journal of
Materials in Civil Engineering* with 56 and 51 articles on the topic respectively.
The subject categories: Construction and Building Technology, and Materials

Science, Multidisciplinary appeared seven times in the top ten active journals focused on recycled concrete. As expected, research about recycled concrete was positioned within the construction technology and material science body of knowledge.

Term analysis

The title and abstract of a journal paper give a concise insight into its content. An overview of the trends in research about recycled concrete emerged from an analysis of the title and abstract of the selected journal papers using VOSviewer. The dataset contained 538 terms that occurred 10 or more times in the titles or abstracts. The terms occurring most frequently were: "concrete aggregate" (548 documents), "construction" (240 documents), and "mixture" (204 documents). To discern the evolution of key issues discussed in research about recycled concrete, a map of terms that occurred in 50 or more publications was created. The terms occurred in the 1,018 journal articles published between 1983 and 2018. Three main clusters were identified.

Owing to the relatedness of the terms, there was an overlap between the clusters at the centre of Figure 13.4. The red cluster represented the properties of recycled concrete. The property of a material is strongly related to the possibility

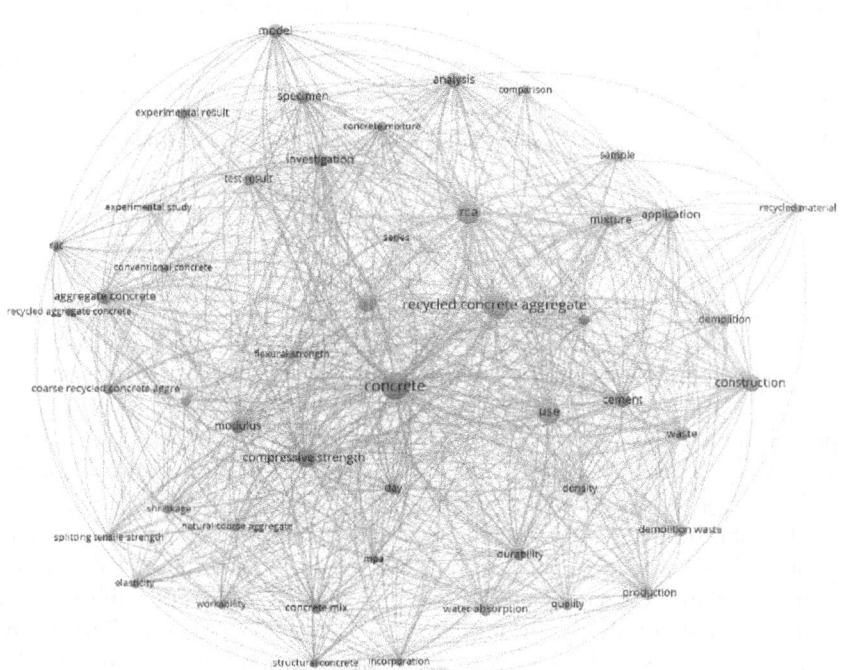

Figure 13.4 Term analysis of research about recycled concrete.

Table 13.5 Occurrence frequency of key terms of time

1983–1994	1995–2006	2007–2018
Cement (4)	Recycled concrete (14)	Recycled concrete (827)
Concrete (4)	Recycled aggregate (11)	Concrete (751)
Calcium silicate (4)	Effect (9)	Compressive strength (356)
Calcium hydroxide (3)	Performance (9)	Aggregate (354)
Recycled concrete (3)	Recycling (6)	Specimen (243)
Fly ash (2)	Time (6)	Demolition (194)
Old concrete (2)	Water-cement ratio (6)	Modulus (171)
Portland cement (2)	Experiment (5)	Experiment (142)
Sand lime (2)	Conventional concrete (4)	Water absorption (122)
Siliceous particle (2)	Flexural strength (4)	Density (106)
	Recycled coarse aggregate (4)	Fly ash (91)
	Shrinkage (4)	Comparison (73)
		Shrinkage (72)

Source: Original.

of its usage in the real world. Generally, the aim of research focused on concrete is to understand its properties, and to explore extant expertise for managing the quantity of cement utilised to achieve reduction of the carbon footprint associated with concrete production and micro-structure (Narayanan and Ramamurthy, 2000). The key terms in the red cluster included modulus, flexural strength, compressive strength, shrinkage, splitting tensile strength, elasticity, and workability.

The green cluster represented the material being investigated and the process of extracting the material. The terms in this cluster included recycled concrete aggregate, demolition, waste, and recycled material. The blue cluster represented the methods for data collection and analysis used in research about recycled concrete. This cluster included terms such as experimental results, experimental study, specimen, and model.

To highlight the trends in research about recycled concrete over time, temporal variation of terms (the content of research papers) can be used to gauge the focus of research in a particular field. The journal articles were grouped into three periods of 12 years: (i) 1983–1994 (five papers); (ii) 1995–2006 (38 papers); and 2007–2018 (975 papers). The frequency of occurrence of terms over these three periods is summarised and presented in Table 13.5. Considering the period between 1983 and 1994, the most frequent terms were cement, concrete, and calcium silicate. Subsequently, the most frequent terms between 1995 and 2006 were recycled concrete, recycled aggregate, effect, and performance. In recent times, the most frequent terms were recycled concrete, concrete, and compressive strength. A scan of the frequently occurring terms in the titles and abstracts of the selected journal papers indicated that the understanding of the properties of recycled concrete has been the focus of a significant portion of studies in this area.

Reflections on the use of bibliometric analysis

From the foregoing, the usefulness of the bibliometric analysis method for appraising the state-of-the-art recycled and self-mitigating research and associated trends is evident. Also, the method proved to be insightful in identifying the productivity of authors and journals focused on this area of research as well as the nature of the collaborative networks existing within the research domain. Such collaborative networks were mainly geographical and institutional. The analysis of the bibliographic information was made simpler by the features of the selected database, SCOPUS, and VOSviewer. The database made it possible to establish patterns in the bibliographic datasets obtained by the search, whereas the VOSviewer proved to be critical to the pictorial representation of these relationship patterns.

Furthermore, key components which are critical of a successful bibliometric analysis were identified. Such components included the use of the following: an incisive research question, an appropriate combination of keywords during the database search, selection of an appropriate database, determination of the unit(s) of analysis in a manner that enables the realisation of the research objective, and the adoption of an effective analytical tool for data analysis and visualisation like the VOSviewer. In the absence of these key components, the possibility of carrying out a successful bibliometric analysis would doubtful.

Conclusion

This chapter contained a report on a bibliometric analysis conducted on the research trends associated with recycled and self-mitigating research about concrete since 1983 until 2018. This report was informed by a need to appraise and showcase the usefulness of the bibliometric analysis method in carrying out built environment and/or construction management research studies. Judging from the successful completion of the study reported in this chapter, it would be an under-statement to say that the bibliometric analysis method offers a different vista to carrying out research within the BE knowledge domain. The efficacy of the method in measuring the productivity of authors working within the research area was noted together with the collaborative networks that exist between these authors. The bibliometric analysis facilitated an assessment of journal productivity, taking into consideration the number of publications contained in the various journals that address the topic of self-mitigating research about recycled concrete.

The use of bibliometric analysis holds salient implications for researchers working within the context of built environment and construction management research clusters, as it assists in identifying the prevailing research gaps in topical areas, whilst making it possible to determine future directions of research about specific phenomena. However, to derive the benefits from the use of this method, it is necessary for the researcher to be skilled in carrying out database searches, as this is critical to sourcing bibliometric information successfully.

References

Allison, D.B., Shiffrin, R.M. and Stodden, V. (2018). Reproducibility of research: Issues and proposed remedies. *Proceedings of the National Academy of Sciences*, 115(11), pp. 2561–2562.

Adams, K.T., Osmani, M., Thorpe, T. and Thornback, J. (2017). Circular economy in construction: Current awareness, challenges, and enablers. *Proceedings of the Institution of Civil Engineers-Waste and Resource Management*, 170(1), pp. 15–24.

Arora, M., Raspall, F., Cheah, L. and Silva, A. (2019). Residential building material stocks and component-level circularity: The case of Singapore. *Journal of Cleaner Production*, 216, pp. 239–248.

Awuzie, B. and Monyane, T.G. (2020). Conceptualizing sustainability governance implementation for infrastructure delivery systems in developing countries: Success factors. *Sustainability*, 12(3), p. 961.

De Bellis, N. (2009). *Bibliometrics and Citation Analysis: From the Science Citation Index to Cybermetrics*, Scarecrow Press, Plymouth.

De Oliveira, R.I., Sousa, S.O. and De Campos, F.C., (2019). Lean manufacturing implementation: Bibliometric analysis 2007–2018. *The International Journal of Advanced Manufacturing Technology*, 101(1–4), pp. 979–988.

Falagas, M.E., Pitsouni, E.I., Malietzis, G.A. and Pappas, G. (2008). Comparison of PubMed, Scopus, Web of Science, and Google Scholar: Strengths and weaknesses. *FASEB Journal*, 22, pp. 338–342.

Geldermans, R.J. (2016). Design for change and circularity – Accommodating circular material and product flows in construction. *Energy Procedia*, 96, pp. 301–311.

Ghaffar, S.H., Burman, M. and Braimah, N. (2020). Pathways to circular construction: An integrated management of construction and demolition waste for resource recovery. *Journal of Cleaner Production*, 244, p. 118710.

Goyal, N. (2017). A review of policy sciences: Bibliometric analysis of authors, references, and topics during 1970–2017. *Policy Sciences*, 50(4), pp. 527–537.

Gradeci, K., Labonnote, N., Time, B. and Köhler, J. (2017). Mould growth criteria and design avoidance approaches in wood-based materials – A systematic review. *Construction and Building Materials*, 150, pp. 77–88.

Heeren, N. and Hellweg, S. (2019). Tracking construction material over space and time: Prospective and geo-referenced modelling of building stocks and construction material flows. *Journal of Industrial Ecology*, 23(1), pp. 253–267.

Malešev, M., Radonjanin, V. and Marinković, S. (2010). Recycled concrete as aggregate for structural concrete production. *Sustainability*, 2(5), pp. 1204–1225.

Martín-Martín, A., Orduna-Malea, E., Ayllón, J.M. and Lopez-Cozar, E.D. (2016). The counting house: Measuring those who count. Presence of bibliometrics, scientometrics, informetrics, webometrics and altmetrics in the Google Scholar citations, Researcherid, ResearchGate, Mendeley and Twitter. arXiv *preprint arXiv*:1602.02412.

McNeil, K. and Kang, T.H.K. (2013). Recycled concrete aggregates: A review. *International Journal of Concrete Structures and Materials*, 7(1), pp. 61–69.

Mingers, J. and Leydesdorff, L. (2015). A review of theory and practice in scientometrics. *European Journal of Operational Research*, 246(1), pp. 1–19.

Mongeon, P. and Paul-Hus, A. (2016). The journal coverage of Web of Science and Scopus: A comparative analysis. *Scientometrics*, 106(1), pp. 213–228.

Narayanan, N. and Ramamurthy, K. (2000). Structure and properties of aerated concrete: A review. *Cement and Concrete Composites*, 22(5), pp. 321–329.

Ogunmakinde, O., Sher, W. and Maund, K. (2017). Circular construction: Opportunities and threats. *2017 Project Management Symposium*, 4–5 May 2017, University of Maryland., College Park, MD, pp. 54–64.

Pan, X., Yan, E., Cui, M. and Hua, W. (2018). Examining the usage, citation, and diffusion patterns of bibliometric mapping software: A comparative study of three tools. *Journal of Informetrics*, 12(2), pp. 481–493.

Perianes-Rodriguez, A., Waltman, L. and van Eck, N.J. (2016). Constructing bibliometric networks: A comparison between full and fractional counting. *Journal of Informetrics*, 10(4), pp. 1178–1195.

Schmidt, F. (2008). Meta-analysis: A constantly evolving research integration tool. *Organizational Research Methods*, 11(1), pp. 96–113.

Sivertsen, G., Rousseau, R. and Zhang, L. 2019. Measuring scientific contributions with modified fractional counting. *Journal of Informetrics*, 13(2), pp. 679–694.

van Eck, N.J. and Waltman, L. (2013). *VOSviewer manual*, Universiteit Leiden, Leiden, vol. 1, issue 1, pp. 1–53.

van Nunen, K., Li, J., Reniers, G. and Ponnet, K. (2018). Bibliometric analysis of safety culture research. *Safety Science*, 108, pp. 248–258.

Wang, K., Vanassche, S., Ribeiro, A., Peters, M. and Oseyran, J. (2017). Business models for building material circularity: Learnings from frontrunner cases. *International HISER Conference on Advances in Recycling and Management of Construction and Demolition Waste*, 21–23 June 2017, Delft University of Technology, Delft, pp. 315–318.

Zupic, I. and Čater, T. (2015). Bibliometric methods in management and organization. *Organizational Research Methods*, 18(3), pp. 429–472.

14 Using literature-based discovery in built environment research

Nathan Kibwami and Apollo Tutesigensi

Introduction

Literature-based discovery (LBD), proposed by Don R. Swanson (Swanson, 1986), is a form of text interrogation of two (or more) juxtaposed, disparate scientific bodies of literature to identify "... non-trivial assertions that are implicit ..." (Smalheiser, 2012, p. 218). LBD involves a systematic, two-component approach to bridging disparate research disciplines, through text mining (Kostoff, 2006, p. 924). The text mining aspect involves extraction and analysis of information expressed in the form of text (Ittipanuvat *et al.*, 2013; Smalheiser, 2012, p. 218; Miyanishi *et al.*, 2010; Kostoff, 2006). If indirect links are found between the disparate literature and no one has previously reported them, new knowledge is created (Weeber *et al.*, 2001). As such, LBD involves a form of syllogism (see Figure 14.1), prescribing that for disparate literature A and C, if A reports a relationship (AB) with a term B, and C reports a relationship (BC) with the same term B, hypotheses (AC) can be derived by connecting A and C (Smalheiser, 2012).

Pioneering the use of LBD is credited to Swanson's medical-related research work published in 1986. From two disparate scientific bodies of literature, one related to fish oil and another to Raynaud's disease, Swanson "proposed [a] hypothesis that fish oil might ameliorate Raynaud's syndrome" (Swanson, 1986, p. 12). This hypothesis was later tested empirically and found acceptable (DiGiacomo *et al.*, 1989). Several subsequent LBD studies in medical disciplines have been undertaken since, resulting in several hypotheses, which have been tested empirically and accepted (Srinivasan, 2004; Weeber *et al.*, 2001; Lindsay and Gordon, 1999).

Outside medical research, LBD has been cited in addressing a variety of research problems (Ittipanuvat *et al.*, 2014; Kostoff *et al.*, 2008b; Cory, 1997), some of which relate to built environment (BE) research (Yung *et al.*, 2013; Dixit *et al.*, 2010). Proponents of LBD in BE research believe that this method has potential in addressing some BE research problems, if applied correctly (Kibwami and Tutesigensi, 2014). These sentiments have been supported in works related to research methods for construction (Fellows and Liu, 2015) and systematic literature reviews related to LBD (Thilakaratne *et al.*, 2019).

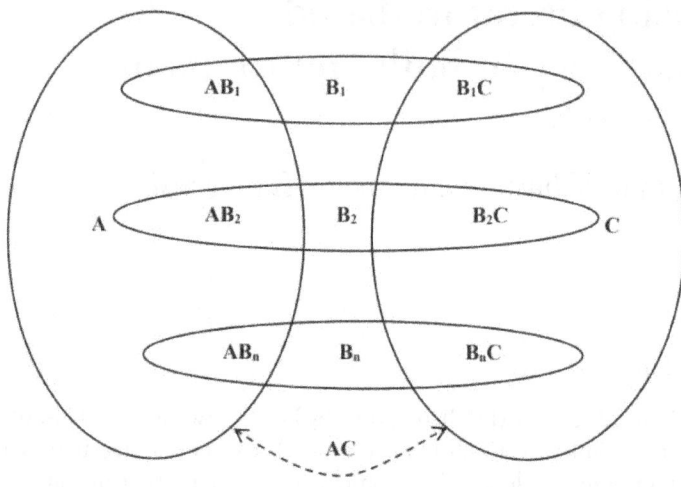

Legend
A - Solution Literature (e.g. literature on fish oils)
C - Problem Literature (e.g. literature on Raynaud's disease)
B - Linking Term (e.g. blood viscocity)
AB - Relationship between A and B (e.g. fish oils reduce blood viscocity)
BC - Relationship between B and C (e.g. Raynaud's patients have high blood viscocity)
AC - Hypothesis (e.g. fish oils can treat Raynaud's disease)

Figure 14.1 The ABC model of LBD; dotted arrow denotes potential hypotheses connecting A and C. (Adapted from Swanson and Smalheiser, 1997, with permission from Elsevier.)

However, the way in which LBD has been used so far in BE studies, suggests violation of its fundamental principles. If such violations are not checked, the purposes for which LBD was developed originally will be lost gradually in BE research. The focus in this chapter is on facilitating robust understanding of LBD among BE researchers in order to improve the accuracy and frequency of LBD in BE research. In subsequent sections of this chapter, the epistemology of LBD and standard approaches to LBD are highlighted before discussing current use of LBD in BE research and associated problems. After this, an approach to facilitate authentic use of LBD in BE research is presented before drawing conclusions.

Epistemology of LBD

The objective of any research is to advance knowledge within the respective field, topic, or context of inquiry. However, there is a possibility that a solution to a problem prevalent in a certain context might be available unknowingly in another disparate field, where researchers are oblivious to the problem (Hristovski *et al.*, 2005). Such inadvertent solutions can remain undiscovered and consequently unpursued, if no inquiry is ever made to consider the disparate fields together. Revelation of such undiscovered public knowledge is the focus of LBD.

The epistemological assumption of LBD is grounded in the idea that there is knowledge hidden in scientific literature. The sheer volume and growth of scientific literature makes it almost impossible for researchers to be aware of or keep up-to-date with advances within, let alone outside, their own research disciplines (Yetisgen-Yildiz and Pratt, 2009; Srinivasan, 2004). Owing to the quantity of information, researcher specialisation is necessary, but this leads to increased fragmentation and thus a vast number of mutually isolated specialities (Ittipanuvat *et al.*, 2014; Swanson, 1991). The increased fragmentation gradually creates an infinite growth of indirect connections amongst specialities, some of which, unbeknown, might offer answers to important prevalent problems (Swanson, 1991).

Therefore, the assumption is made in LBD that the sum of the world's knowledge is greater than the sum of knowledge embedded in scientific literature (Cory, 1997). Through a process of "mining" scientific literature, these implicit linkages carrying implicit knowledge can be revealed, which can lead to the creation of new knowledge (Lekka *et al.*, 2011).

Approaches to LBD

There are two fundamental approaches to LBD: open discovery (OD) and closed discovery (CD) (Kostoff *et al.*, 2008a; Weeber *et al.*, 2001).

Open discovery

In OD, the process starts with literature about the problem, the "problem literature" (C), and using a correlation mining technique such as "linking-term count" (Yetisgen-Yildiz and Pratt, 2006) and "linking terms" (B) related to the problem are identified, and then through a similar technique, "target terms" related to the linking terms are identified in disparate "solution literature" (A) (Yetisgen-Yildiz and Pratt, 2009; Weeber *et al.*, 2001) (see Figure 14.2). Hypotheses are then derived by connecting the problem literature C and solution literature A. In his early work, Swanson (1986) referred to OD as Procedure 1 of LBD (Swanson and Smalheiser, 1997), while others referred to it as "one node" LBD (Smalheiser *et al.*, 2009) or one-directional procedure (Srinivasan, 2004) characterised by the generation of hypotheses (Weeber *et al.*, 2001).

Closed discovery

CD starts with the problem literature (C) and solution literature (A) simultaneously, as illustrated in Figure 14.3. The assumption of CD is that some connections between A and C are already known, either derived from the OD process or stated by conjecture, and therefore, the aim of CD is to identify new connections (Srinivasan, 2004). Consequently, common linking B-terms are identified, working towards identifying new linkages between the two bodies of literature (Kostoff *et al.*, 2008a). Various terminologies are used often to refer to the CD process, e.g. Procedure 2 of LBD (Swanson and Smalheiser, 1997) and the two-node approach

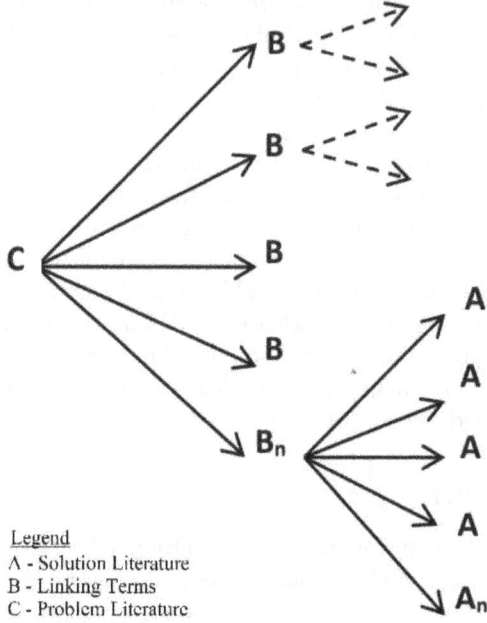

Figure 14.2 The open discovery LBD process; dotted lines denote unsuccessful pursuits. (Adapted from Weeber *et al.*, 2001, with permission from John Wiley and Sons.)

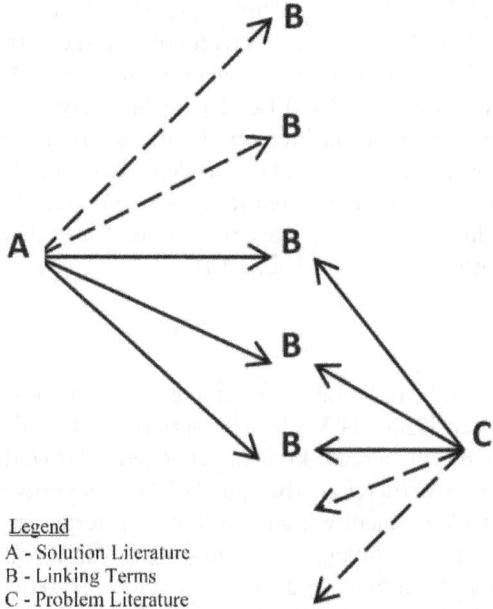

Figure 14.3 The closed discovery LBD process; dotted lines denote unsuccessful pursuits. (Adapted from Weeber *et al.*, 2001, with permission from John Wiley and Sons.)

(Smalheiser *et al.*, 2009). CD primarily offers a mechanism of testing or confirming hypotheses (Weeber *et al.*, 2001), but in the process, there is potential to generate new hypotheses (Smalheiser *et al.*, 2009).

CD is the most predominant approach to LBD and was used in the illustrative example presented in this chapter. Indeed, some researchers (e.g. Smalheiser *et al.*, 2009) have developed tools such as Arrowsmith that provide guidance on carrying out CD (Arrowsmith, 2007). Arrowsmith uses sophisticated text mining algorithms, analyses disparate literature, and returns potential B-terms that the investigator can follow up to discover novel linkages.

A critique of LBD in BE research

Outside medical research, LBD has been cited in a number of studies addressing a variety of research problems, including some in BE research (Yung *et al.*, 2013, Dixit *et al.*, 2010).

In a study to identify parameters that lead to differing embodied energy measurements among buildings, Dixit *et al.* used LBD as the "research method" (Dixit *et al.*, 2010). Dixit *et al.* (2010) stated that a concept of triangulation, involving "cross-referencing various sources of information about the same phenomenon", was used to identify parameters that caused variation in embodied energy figures of buildings. It was reported in the study that a literature search revealed 10 parameters that influenced embodied energy measurements. These results were presented in the form of a matrix showing the ten parameters identified and the 23 respective literature sources underpinning the findings. From the study, it was concluded that addressing the identified parameters could lead to consistent measurement of embodied energy of building materials. However, no explicit conclusions were given on the efficacy of using LBD. A similar approach of using LBD was used in another related study by Dixit *et al.* (2013).

In a study seeking to provide an audit of life cycle energy analysis of buildings, it was stated that "a literature-based discovery method has been adopted" (Yung *et al.*, 2013). As justification of the research method, it was stated in the study that, citing Dixit *et al.* (2010), LBD had been applied previously to energy studies of buildings. In the study, a literature review on life cycle energy studies of buildings was done, consequently identifying 38 research works, consisting of 206 cases (i.e. buildings), across 16 countries. The embodied energy in each of the cases was identified and summaries were presented in scatter plots showing differences in the total embodied and operational energy of various types of buildings. From the study, it was concluded that by using LBD, a database of life cycle energy of buildings was created.

Based on evidence gathered from medical LBD studies, where LBD originated, LBD requires disparate literature in more than one context. A literature search must be performed on two disparate contexts. However, literature from a single context was used in the two previous BE studies (Yung *et al.*, 2013, Dixit *et al.*, 2010). The literature considered in Dixit *et al.* (2010) was limited to embodied energy analysis, while that in Yung *et al.* (2013) was limited to life cycle energy analysis of buildings. Using literature from a single context does not facilitate

LBD, since it is expected that the solution or problem would have been discovered or well known within that context (Kostoff *et al.*, 2008a). In addition, the literature used in these BE studies was not disparate, since it had several cross-citations. For instance, an article (Menzies *et al.*, 2007) was cited in another (Hammond and Jones, 2008), yet both were used in Dixit *et al.* (2010). Similarly, in Yung *et al.* (2013), articles (e.g. Monahan and Powell, 2011; Asif *et al.*, 2007) are used, yet one of them cites the other. By violating LBD's fundamental principle of "using multiple disparate literature", the claim in these BE studies to have used LBD was not well founded.

Studies outside BE in which LBD was used (Ittipanuvat *et al.*, 2014; Kostoff *et al.*, 2008b; Cory, 1997) provided clear articulation of the application of authentic LBD. In such studies, features of LBD (e.g. linking of terms, hypotheses) were articulated explicitly. However, besides merely citing that LBD literature databases were used, there was no explicit articulation of how and why LBD was used in the BE studies (Dixit *et al.*, 2013; Yung *et al.*, 2013; Dixit *et al.*, 2010). While using LBD was acknowledged in Dixit *et al.* (2010), nothing was mentioned about how LBD was suited to the aim of the work. Similarly, in Yung *et al.* (2013, p. 45), there were no explanations as to why LBD was appropriate to use other than claiming that it "matches perfectly the aim of [that] paper". Since there was little evidence and details to confirm that LBD featured in addressing the problem, it is difficult to conclude that LBD was appropriate for the research questions addressed in these BE studies.

Furthermore, confusing LBD with systematic literature review was evident, since barely any features of LBD could be traced in these BE studies. Hence, there is little evidence to reject a hypothesis that systematic literature reviews (and not LBD) were carried out. For instance, their results do not provide any linkages or hypotheses typical of LBD findings, but instead were summarised in a fashion typical of systematic literature reviews. Certainly, these BE studies were not different from other similar literature review studies (e.g. Ibn-Mohammed *et al.*, 2013; Dakwale *et al.*, 2011; Ramesh *et al.*, 2010; Menzies *et al.*, 2007; Casals, 2006) that did not mention LBD.

Overall, it appears that some BE researchers assume LBD to be another form of literature review, a situation which potentially propagates confusion with respect to the differences between LBD and literature review. LBD and literature review are not the same and, if LBD BE studies do not articulate this, confusion of LBD with literature review will persist. This unfortunately will hinder full exploitation of the potential of LBD in BE research. This situation must be avoided. This was the motivation for providing guidance for BE researchers using LBD, in the section below.

The proposed LBD approach

Following a CD process, an LBD approach composed of five steps is proposed in this section. Being a CD-based LBD approach, it is presupposed that the researcher has specified the two disparate literature contexts, with some preliminary propositions to pursue the investigation.

Step 1: Literature data retrieval

A comprehensive literature search is performed on two identified, disparate contexts of inquiry (i.e. A and C) to generate the *corpora* for performing LBD; preference should be given to peer-reviewed journal articles. This step is similar in many ways to that of systematic searches performed in conducting systematic literature reviews.

Step 2: Term extraction

It is necessary to consider both "recall" and "precision" in the procedure used to extract terms (Naumann and Herschel, 2010). "Recall" relates to the number of terms that can be retrieved, whereas "precision" is related to the relevance or plausibility of the extracted terms. Higher precision can be guaranteed only at the expense of lower recall and vice versa (Ganti and Sarma, 2013). Therefore, term extraction procedures (or software packages) that rely solely on statistical information (e.g. frequency of occurrence) are not preferable, since many implausible terms (e.g. is, of, the) will be extracted. An approach that can balance both precision (based on linguistics) and recall (based on statistics) is necessary. The C-value/NC-value method of text mining (Frantzi *et al.*, 1998) is recommended, since both statistical and linguistic information are considered.

On statistical information

Term extraction in LBD involves using statistical procedures (Yetisgen-Yildiz and Pratt, 2006; Frantzi *et al.*, 1998). However, usually in these statistical procedures, a particular string of characters is treated solely as an instance of a word or phrase without reference to its deeper linguistic significance (Lindsay and Gordon, 1999, p. 575), in a similar way to text extraction processes prevalent in other techniques of textual analysis, such as manifest content analysis (Hsieh and Shannon, 2005). The LBD term-extraction statistics commonly used, and included in the suggested approach are: term frequency (Tf), document frequency (Df), inverse document frequency (iDf), and term frequency – inverse document frequency (Tf-iDf) (Lindsay and Gordon, 1999, Ittipanuvat *et al.*, 2014, Frantzi *et al.*, 1998, Gordon *et al.*, 2002).

Initially, terms should be sorted or ranked according to their Tf (i.e. number of times a term appears in a *corpus*). However, using Tf alone for further evaluation means that terms appearing less frequently might be missed out (i.e. because they are ranked low), yet they might be plausible. To circumvent this, the concept of iDf developed in Jones (1972) is suggested. The iDf weighting boosts terms with low frequency, yet are concentrated in few specific documents/articles. The iDf is computed as log (D/Df), where D is the total number of documents in the *corpus* being considered and Df is the number of documents in which a term appears. This consequently yields a Tf-iDf measure, which is the product of Tf and iDf (Salton and Buckley, 1988). Tf-iDf is usually a preferred measure and has been cited in several LBD studies as a better measure of the relevance of a term than Tf

(Ittipanuvat *et al.*, 2013; Srinivasan, 2004; Lindsay and Gordon, 1999). Therefore, terms should be ranked by Tf-iDf and low-ranking terms may be discarded.

On specifying linguistic information

Without applying any linguistic information in term extraction, statistical measures alone can generate many terms, most of which might be implausible (i.e. with low precision). However, using linguistic information, such as tagging, linguistic filtering, and stop word list (see Frantzi *et al.*, 1998), improves the precision of terms to be extracted from each context. In applying linguistic information, tagging is used to attach grammatical tags (e.g. noun, adjective) to each term in the *corpus*. Then the appropriate linguistic filter applied uses the grammatical tags to extract specified terms (e.g. nouns only, verbs only, adjectives only). Illustrations of linguistic filtering reveal that most terms are composed usually of nouns, verbs, or adjectives, and multi-word terms usually consist of at least a noun (Frantzi *et al.*, 1998). Thus, terms extracted should be linguistically filtered for nouns, verbs, and adjectives in that order of preference. For automated filters such as those in medical databases, linguistic filtering can be set automatically. The approach suggested in the example used in this case is semi-automated, since linguistic filtering is done manually by inspecting each extracted term. The stop word list, which has been used previously in many LBD studies (Lindsay and Gordon, 1999; Swanson and Smalheiser, 1997), is then used to distinguish between potentially useful and non-useful terms, such as those that frequently occur (e.g. is, to, what) (Weeber *et al.*, 2001). The stop list can be pre-compiled based on the predicted suitability of terms (Swanson and Smalheiser, 1997) or compiled concurrently with the term extraction process (Lindsay and Gordon, 1999). As part of the synonymy and stemming rules (Lindsay and Gordon, 1999), it is suggested that only exactly matching words (i.e. not synonyms) should be considered in order to control unnecessary recall and noise. However, singular-plural stemming rules (Lindsay and Gordon, 1999) can be applied and, in such cases, the terms (e.g. house and houses) should be combined into one.

Depending on the procedure used for extracting terms (i.e. manual, automatic), a manageable number and length of terms should be considered. In well-structured and online *corpora* (e.g. in MEDLINE), it is possible to know the approximate number of terms with which to work (Weeber *et al.*, 2001, p. 551). However, for a semi-automated process, as suggested in this case, where articles are gathered manually from different databases, only an estimate is possible. For instance, for literature consisting of 20 articles, assuming an average length of a full article to be 7,000 words, this would mean working with 140,000 terms. To manage the winnowing process towards precision, an initial working number of terms from each context should be set. Meanwhile, the decision of setting the minimum length of terms (i.e. number of characters per term) depends on the desired precision and recall. Shorter terms are better in terms of recall but not precision. Also, terms can be unigrams (i.e. one word terms), bigrams (i.e. two-word terms), or n-grams (Ittipanuvat *et al.*, 2013; Frantzi *et al.*, 1998). For the current approach, unigrams were considered because of some limitations highlighted later.

The recall for unigrams is usually high, since unigrams can exist either on their own or as nested terms (i.e. sub-terms of bigrams or n-grams). In Ittipanuvat *et al.* (2014), unigrams accounted for over three quarters of the total terms extracted.

Step 3: Category development

Dissimilar to biomedical databases, where terms can be classified automatically into their respective pre-determined semantic categories (Smalheiser *et al.*, 2009), manual categorisation is suggested, which demands human intervention and acquaintance with qualitative data analysis techniques, such as latent content analysis (Hsieh and Shannon, 2005) and the paradigm model (Strauss and Corbin, 1998). However, the intensity of human intervention does not manifest entirely as a disadvantage since it means the analyst must be familiar with what the literature is saying. To guide the categorisation process, a paradigm model, initially proposed in Strauss and Corbin (1998) and subsequent texts (Corbin and Strauss, 2008), is suggested. It consists of phenomena (i.e. what is going on?), conditions (i.e. what are the causes?), actions/interactions (i.e. what is the response?), and consequences (i.e. what are the results?). Following such questions posed in the components of the paradigm model, for each of the key terms, the literature where it appears is read by line, paragraph, page, or entire article, in order to elucidate the context of how the term was used. Essentially, the approach involves "coding", which is "… an analytic process through which data are fractured, conceptualised, and integrated …" (Strauss and Corbin, 1998, p. 3). Coding is done per sentence and paragraph following the key search terms only, and appropriate software can be used to aid the process. Consequently, categories are developed from the key terms and it is possible for a given term to belong to several categories.

Step 4: Semantic similarity

A semantic similarity measure is a "… function that, given two … sets of terms annotating two entries, returns a numerical value reflecting the closeness in meaning between them" (Pesquita *et al.*, 2009, p. 1). Several similarity measures (e.g. Jaccard Index, Dice coefficient, and cosine) are discussed in the literature (Ganti and Sarma, 2013; Naumann and Herschel, 2010; Pesquita *et al.*, 2009), which also are used often in LBD studies (Ittipanuvat *et al.*, 2013; Miyanishi *et al.*, 2010). In this case, the cosine similarity measure was suggested. At this stage, categories based on the key terms only would have been developed from the *corpora*. From appropriate coding software, it is possible to map the extracted terms that intersect with a given category and therefore, several term combinations (e.g. terms in both A and C, in A only, and in C only) associated with a developed category can be worked out. Like the key terms, it is possible for a given extracted term to belong in several categories.

The categories are then transformed into vectors and the similarity between them is computed. Since vectors only work with integers, each term is therefore represented by its Tf-iDf measure. Put another way, a category composed of terms is represented as a vector composed of Tf-iDfs. This idea, suggested initially in

Salton and Buckley (1988), is usually used in works related to textual analyses, document indexing, and document retrieval. The similarity between two vectors is a property of the cosine of the angle between them (i.e. 1 if the vectors are identical and zero if they are not). The cosine values are computed using the cosine vector similarity formula (Salton and Buckley, 1988, p. 514) as per Equation (14.1):

$$\text{Similarity } (A_v, C_v) = \sum(w_{A,t} \times w_{C,t}) \Big/ \left(\sqrt{\sum w_{A,t}^2} \times \sqrt{\sum w_{C,t}^2} \right) \tag{14.1}$$

where A_v and C_v are Tf-iDf vectors representing literature contexts of A and C respectively; $w_{A,t}$ and $w_{C,t}$ are weights (i.e. Tf-iDf) of a term t with regard to literature A and C respectively.

Step 5: Deducing relationships

This step is the climax of a typical LBD study and involves deduction of relationships (i.e. plausible hypotheses) or confirmation of the same. In this case, deducing relationships was based on the cosine similarity measure and the Tf-iDf measure. It was assumed that vectors (i.e. categories) with cosine similarity values closer to 1 would be more closely related and thus suitable for generating plausible hypotheses. This assumption was not new (see Miyanishi et al., 2010, p. 1554). However it should not be taken as the only guidance in this step because although lower cosine values suggest remote relationships between categories, these relationships may, *ipso facto*, be sources of new knowledge. Nonetheless, the key guidance to pursue any plausible hypothesis/relationship regarding any term in the vectors is based on the cosine similarity score and the term's rank/weighting (i.e. Tf-iDf). In other words, it is inferred that the plausibility of a hypothesis linking an A-Term to a C-Term is related to the cosine similarity between the two vectors that describe how those terms manifest in A and C.

Application of the proposed LBD approach

Table 14.1 provides a step-by-step summary of the outcomes of the proposed LBD approach that was applied in 2013 with the aim of identifying lessons about managing carbon emissions that Uganda could learn from the United Kingdom. The project was based on the idea that the disparity of literature necessary for LBD application can be provided by country context (Gordon and Awad, 2008). Personal experience and anecdotal evidence suggested that there was little, if anything, implemented in the Ugandan building sector to address carbon emissions. This was confirmed by a nil return when a systematic search for literature was implemented using the key words of "building(s)" or "construction" and "carbon emissions". A similar search involving the UK returned a rich collection of publications. One of the testable hypotheses that emerged was about introducing embodied carbon emissions of buildings in Uganda into the United Nations Framework Convention on Climate Change's Clean Development Mechanism (CDM). Following this, a CDM project was proposed (Kibwami and Tutesigensi, 2015) and may be tested in the future.

Table 14.1 Illustration of the new LBD approach

Step	Action and outcomes
1. Literature data retrieval	A comprehensive literature search about carbon emissions in Uganda (comprising problem literature C) and carbon emissions in the United Kingdom (comprising solution literature A) was undertaken; a total of 105 articles were identified (29 and 76 for C and A respectively)
2. Term extraction	To balance precision and recall, while presenting a manageable number of terms, 1,000 terms were extracted and ranked from each context. A number of terms present in both A and C (i.e. the B-terms), including buildings, climate, renewable, energy, costs, and technology emerged
3. Category development	The most prominent categories identified and considered were strategies to address emissions, causes of emissions, barriers to reducing emissions, and regulations related to emissions. Terms that belonged to each of the categories, with respect to A and C, were extracted and ranked by their Tf-iDf measure
4. Semantic similarity	The cosine similarity of the category "strategies to address emissions" between A and C in relation to B terms was 0.8616 (see Appendix A). This result implied a good relationship between A and C, suggesting that perhaps similar initiatives of addressing emissions existed in both contexts. However, when terms not common to both contexts were included in the computations, the similarity reduced to 0.3005 (see Appendix B). This reduction demonstrated that the manifestation of "strategies to address emissions" in the two contexts was different and there could be ideas about addressing carbon emissions in context A which could benefit context C
5. Deducing relationships	Upon critical consideration of the A only and C only literature, plausible hypotheses were posed regarding addressing emissions associated with buildings in context A. One such hypothesis was as follows: CDM (clean development mechanism) will reduce embodied carbon emissions of *residential* buildings in Uganda

Source: Original.

Conclusions

In this chapter, it was demonstrated that the usefulness of LBD lies in its capacity to help researchers generate hypotheses about a given problem by deducing connections between bodies of literature from two disparate contexts. The authenticity of LBD application should always be judged against this principle.

Although LBD is gaining increasing recognition, it is still limited mostly to addressing medical problems. Nonetheless, its uptake in BE research appears to be emerging, albeit with some deficiencies. Thus far, LBD in BE research has been characterised by misconception, which has manifested in the form of focusing on single instead of two or more disparate bodies of literatures; insufficient consideration to justify the choice of method; and confusing LBD with mainstream systematic literature review. There is an urgent need to move from this *status quo* towards authentic application of LBD in BE research.

To facilitate the move to authentic application of LBD, a five-step approach to LBD was demonstrated, which adheres to the fundamental principles of LBD and its application was illustrated with an example. The approach involves literature data retrieval, term extraction, category development, determination of semantic similarity, and deduction of relationships. The proposed approach demonstrates robust use of qualitative information to derive quantitative indicators that enable researchers to link phenomena that appear in disparate contexts and propose testable hypotheses. BE researchers are strongly urged to tap into the demonstrated potential of LBD, whilst adhering to the principles and assumptions of the method.

While the efficacy of LBD has been argued and emphasised in this chapter, it would not be complete without underscoring the associated limitations in BE research. Unlike in biomedical research, the approach presented in this chapter was semi-automated, involving a considerable effort in human assessment, implying that some tasks were limited only to what could be managed reasonably. As such, there could be potential for bias since human perceptions and opinions differ. However, any of these biases will be revealed when the hypothesis testing takes place; so, instead of this being a cause for concern, it should be seen as providing opportunity to discard fantasies. Researchers must, however, be vigilant to avoid missing hypotheses because of researcher bias.

References

Arrowsmith. (2007). *Arrowsmith project: linking documents, disciplines, investigators and databases*. Available at: http://goo.gl/JQe6X6 [Accessed on: 10 November 2019].

Asif, M., Muneer, T. and Kelley, R. (2007). Life cycle assessment: A case study of a dwelling home in Scotland. *Building and Environment*, 42, pp. 1391–1394.

Casals, X.G. (2006). Analysis of building energy regulation and certification in Europe: Their role, limitations and differences. *Energy and Buildings*, 38, pp. 381–392.

Corbin, J.M. and Strauss, A.L. (2008). *Basics of Qualitative Research: Techniques and Procedures for Developing Grounded Theory*, Sage, Los Angeles, CA, London.

Cory, K. (1997). Discovering hidden analogies in an online humanities database. *Computers and the Humanities*, 31, pp. 1–12.

Dakwale, V.A., Ralegaonkar, R.V. and Mandavgane, S. (2011). Improving environmental performance of building through increased energy efficiency: A review. *Sustainable Cities and Society*, 1, pp. 211–218.

Digiacomo, R.A., Kremer, J.M. and Shah, D.M. (1989). Fish-oil dietary supplementation in patients with Raynaud's phenomenon: a double-blind, controlled, prospective study. *American Journal of Medicine*, 86, pp. 158–164.

Dixit, M.K., Culp, C.H. and Fernández-Solís, J.L. (2013). System boundary for embodied energy in buildings: A conceptual model for definition. *Renewable and Sustainable Energy Reviews*, 21, pp. 153–164.

Dixit, M.K., Fernández-Solís, J.L., Lavy, S. and Culp, C.H. (2010). Identification of parameters for embodied energy measurement: A literature review. *Energy and Buildings*, 42, pp. 1238–1247.

Fellows, R.F. and Liu, A. (2015). *Research Methods for Construction*, Wiley-Blackwell, Oxford.

Frantzi, K., Ananiadou, S. and Tsujii, J. 1998. The C-value/NC-value method of automatic recognition for multi-word terms. *Research and Advanced Technology for Digital Libraries*, Springer, Berlin, Heidelberg.

Ganti, V. and Sarma, A.D. (2013). Data cleaning: A practical perspective. *Synthesis Lectures on Data Management*, 5, pp. 1–85.

Gordon, M., Lindsay, R.K. and Fan, W. (2002). Literature-based discovery on the World Wide Web. *ACM Transactions on Internet Technology*, 2, pp. 261–275.

Gordon, M.D. and Awad, N.F. (2008). The tip of the iceberg: The quest for innovation at the base of the pyramid. In: P. Bruza and M. Weeber (eds.), *Literature-Based Discovery*, Springer, Berlin, Heidelberg, vol. 15, pp. 23–37.

Hammond, G.P. and Jones, C.I. (2008). Embodied energy and carbon in construction materials. *Proceedings of the ICE – Energy*, 161, pp. 87–98.

Hristovski, D., Peterlin, B., Mitchell, J.A. and Humphrey, S.M. (2005). Using literature-based discovery to identify disease candidate genes. *International Journal of Medical Informatics*, 74, pp. 289–298.

Hsieh, H.F. and Shannon, S.E. (2005). Three approaches to qualitative content analysis. *Qualitative Health Research*, 15, pp. 1277–1288.

Ibn-Mohammed, T., Greenough, R., Taylor, S., Ozawa-Meida, L. and Acquaye, A. (2013). Operational vs. embodied emissions in buildings – A review of current trends. *Energy and Buildings*, 66, pp. 232–245.

Ittipanuvat, V., Fujita, K., Sakata, I. and Kajikawa, Y. (2013). Finding linkage between technology and social issue: A literature-based discovery approach. *Journal of Engineering and Technology Management*. doi:10.1016/j.jengtecman.2013.05.006.

Ittipanuvat, V., Fujita, K., Sakata, I. and Kajikawa, Y. (2014). Finding linkage between technology and social issue: A literature based discovery approach. *Journal of Engineering and Technology Management*, 32, pp. 160–184.

Jones, K.S. (1972). A statistical interpretation of term specificity and its application in retrieval. *Journal of Documentation*, 28, pp. 11–21.

Kibwami, N. and Tutesigensi, A. (2014). Using the literature based discovery research method in a context of built environment research. In: A.B. Raiden and E. Aboagye-Nimo (eds.), *Proceedings of 30th Annual ARCOM Conference*, 1–3 September 2014, Association of Researchers in Construction Management, Portsmouth, pp. 227–236.

Kibwami, N. and Tutesigensi, A. (2015). Integrating clean development mechanism into the development approval process of buildings: A case of urban housing in Uganda. *Habitat International*, 53, pp. 331–341.

Kostoff, R.N. (2006). Systematic acceleration of radical discovery and innovation in science and technology. *Technological Forecasting and Social Change*, 73, pp. 923–936.

Kostoff, R.N., Briggs, M.B., Solka, J.L. and Rushenberg, R.L. (2008a). Literature-related discovery (LRD): Methodology. *Technological Forecasting and Social Change*, 75, pp. 186–202.

Kostoff, R.N., Solka, J.L., Rushenberg, R.L. and Wyatt, J.A. (2008b). Literature-related discovery (LRD): Water purification. *Technological Forecasting and Social Change*, 75, pp. 256–275.

Lekka, E., Deftereos, S.N., Persidis, A., Persidis, A. and Andronis, C. (2011). Literature analysis for systematic drug repurposing: A case study from Biovista. *Drug Discovery Today: Therapeutic Strategies*, 8, pp. 103–108.

Lindsay, R.K. and Gordon, M.D. (1999). Literature-based discovery by lexical statistics. *Journal of the American Society for Information Science*, 50, pp. 574–587.

Menzies, G.F., Turan, S. and Banfill, P.F.G. 2007. Life-cycle assessment and embodied energy: A review. *Proceedings of the ICE – Construction Materials*, p. 160. Available at: http://www.icevirtuallibrary.com/content/article/10.1680/coma.2007.160.4.135.

Miyanishi, T., Seki, K. and Uehara, K. (2010). Hypothesis generation and ranking based on event similarities. *Proceedings of the 2010 ACM Symposium on Applied Computing*, ACM, Sierre, Switzerland, pp. 1552–1558.

Monahan, J. and Powell, J.C. (2011). An embodied carbon and energy analysis of modern methods of construction in housing: A case study using a lifecycle assessment framework. *Energy and Buildings*, 43, pp. 179–188.

Naumann, F. and Herschel, M. (2010). An introduction to duplicate detection. *Synthesis Lectures on Data Management*, 2, pp. 1–87.

Pesquita, C., Faria, D., Falcao, A.O., Lord, P. and Couto, F.M. (2009). Semantic similarity in biomedical ontologies. *PLOS Computational Biology*, 5(7), pp. 1–12.

Ramesh, T., Prakash, R. and Shukla, K.K. (2010). Life cycle energy analysis of buildings: An overview. *Energy and Buildings*, 42, pp. 1592–1600.

Salton, G. and Buckley, C. (1988). Term-weighting approaches in automatic text retrieaval. *Information Processing and Management*, 24, pp. 513–523.

Smalheiser, N.R. (2012). Literature-based discovery: Beyond the ABCs. *Journal of the American Society for Information Science and Technology*, 63, pp. 218–224.

Smalheiser, N.R., Torvik, V.I. and Zhou, W. (2009). Arrowsmith two-node search interface: A tutorial on finding meaningful links between two disparate sets of articles in MEDLINE. *Computer Methods and Programs in Biomedicine*, 94, pp. 190–197.

Srinivasan, P. (2004). Text mining: Generating hypotheses from MEDLINE. *Journal of the American Society for Information Science and Technology*, 55, pp. 396–413.

Strauss, A.L. and Corbin, J.M. (1998). *Basics of Qualitative Research: Techniques and Procedures for Developing Grounded Theory*, Sage, Thousand Oaks, CA.

Swanson, D.R. (1986). Fish oil, Raynaud's syndrome, and undiscovered public knowledge. *Perspectives in Biology and Medicine*, 30, pp. 7–18.

Swanson, D.R. (1991). Complementary structures in disjoint science literatures. *Proceedings of the 14th Annual International ACM SIGIR Conference on Research and Development in Information Retrieval*, ACM, Chicago, IL, pp. 280–289.

Swanson, D.R. and Smalheiser, N.R. (1997). An interactive system for finding complementary literatures: A stimulus to scientific discovery. *Artificial Intelligence*, 91, pp. 183–203.

Thilakaratne, M., Falkner, K. and Atapattu, T. (2019). A systematic review on literature-based discovery: General overview, methodology, & statistical analysis. *ACM Computing Surveys*, 52, pp. 1–34.

Weeber, M., Klein, H., De Jong-Van Den Berg, L.T.W. and Vos, R. (2001). Using concepts in literature-based discovery: Simulating Swanson's Raynaud–fish oil and migraine-magnesium discoveries. *Journal of the American Society for Information Science and Technology*, 52, pp. 548–557.

Yetisgen-Yildiz, M. and Pratt, W. (2006). Using statistical and knowledge-based approaches for literature-based discovery. *Journal of Biomedical Informatics*, 39, 600–611.

Yetisgen-Yildiz, M. and Pratt, W. (2009). A new evaluation methodology for literature-based discovery systems. *Journal of Biomedical Informatics*, 42, pp. 633–643.

Yung, P., Lam, K.C. and Yu, C. (2013). An audit of life cycle energy analyses of buildings. *Habitat International*, 39, pp. 43–54.

15 Combining study findings by using multiple literature review techniques and meta-analysis

A mixed-methods approach

Samantha Low-Choy, Fernando Almeida, and Judy Rose

Introduction

A mixed-methods research approach is defined as a procedure for using a combination of quantitative and qualitative methods for the collection and analysis of data in the same research study (Creswell and Clark, 2011). However, a multi-method approach is differentiated from a mixed-methods approach in *not requiring both* quantitative and qualitative methods. Instead "multi-method" refers to the use of *multiple* methods that are *either* qualitative or quantitative. A mixed-methods approach is justified when the interaction between quantitative and qualitative methods provides better analytical possibilities than either method alone. Such an approach becomes relevant when researchers intend to perform comparative analyses simultaneously and develop aspects of the study in comprehensive and in-depth terms. However, although mixed methods have gained visibility in recent years, it is necessary that care is taken to avoid methodological and design problems, with over-simplification being a common criticism (Cheek, 2015, p. 630). Rigour is required when integrating evidence within qualitative and quantitative modalities, or traversing the boundaries that separate them. This occurs in studies where the strength of a quantitative analysis, the purpose of which is specifically to confirm effects, is combined with the deep explanatory descriptions obtained from qualitative analyses (Castro *et al.*, 2010).

In this chapter, meta-analysis (MA) is regarded as a mixed method to combine the qualitative component of literature review (LR) to establish a context in which to use quantitative analysis to combine findings, expressed as comparable quantities, across multiple studies. The LR provides a conceptual basis for ensuring that all studies included are comparable, as are the quantities (Paré et al., 2015). In addition, the use of qualitative information might modify the contribution of each study to the analysis. Studies may be re-weighted by sample size or other measures of quality, such as the extent of important sources of bias, for instance, the use of expert assessments. Important aims of MA include: accumulating knowledge; highlighting consistency amongst studies; detecting potential publication bias; and revealing points of consensus or dissent across such studies. In this way, the conclusions produced by MA can be stronger than the findings of any individual study, owing to: the increased amount of evidence, the ability to account for variability between studies as well as within studies, and the cumulative view of outcomes (Cooper *et al.*, 2019). These statistical benefits can be boosted when LR is "mixed in" as a qualitative research method within the MA, including benefits such as: characterisation of the sampling frame, quantities of interest, and of their quality. In this sense, the purpose of this study was to explore the process of mixing LR methods into MA from two fundamental perspectives. First, it is essential to give a theoretical description of this process. It is not always easy or possible to mix both approaches. Therefore, it is important to explore how mixing methods affects the details of implementation for both the LR process and the statistical modelling within a MA. Second, these issues are illustrated in this chapter by using two scenarios in different fields (natural sciences *versus* business/IT) to demonstrate how this process could be applied, whilst identifying useful practices, difficulties, and challenges.

The aim of this study was to provide a fundamental guide for researchers who seek to integrate LR methodologies into a MA using a mixed-methods approach. This will help a researcher to understand the approach, learn how to apply this research methodology, and assess its robustness, challenges, and difficulties during the implementation process.

Quality appraisal using AMSTAR2

Essentially MA begins with LR, potentially modified by experts, to locate relevant studies and extract relevant information about each study, which is then collated across studies using statistical analysis. Criteria are available not only to help guide researchers in good practices for MA, but also to list minimal key requirements of MA. Many criteria are based on the assumption that a systematic LR has been undertaken. AMSTAR2 (Shea *et al.*, 2017) is an update to the original AMSTAR, an instrument widely used to appraise the quality of published, systematic reviews for MA (Farrah *et al.*, 2019). The updated AMSTAR2 criteria can be used to assess both randomised studies (prevalent in medical research) as well as observational and other non-randomised studies (more common in social science). Seven critical domains that broadly correspond to stages of MA are addressed by AMSTAR2 in different items, respectively, as follows:

1   Protocol is stipulated before commencing the literature search (AMSTAR2 Item 2).
2   The literature search is adequate (Item 4).
3   Justification is provided for excluding particular studies (Items 3, 5, 7).
4   Quantities extracted for MA are comparable and adjusted for biases in individual studies (Items 6, 9, 10, 14).
5   Statistical methods of MA are appropriate (Item 11).
6   Interpreting the results accounts for risks of bias (Items 12, 13, 16).
7   Presence and likely impact of publication bias is assessed (Item 15).

Meta-analysis methodology

Typically, MA is presumed to require a systematic review, as reflected in the appraisal criteria preceding AMSTAR2 (above). However, such criteria have evolved primarily in disciplines such as health or medical sciences, where search terms (such as "medical condition" or "procedure") tend to be well defined. In contrast, such searches are extremely challenging in many other fields of enquiry, particularly for the first MA of a topic. In these situations, a systematic review is not necessarily sufficient: it may return a list of studies that is too large (and prohibitive to read to assess eligibility) or too narrow. However, many different methodologies for LR exist. In particular, six of these are relevant to five of the seven domains of MA within AMSTAR2 criteria, listed above. The authors, therefore, propose a description of MA, with stages that are aligned with seven AMSTAR2 domains, and which allows for different LR methodologies (italicised below), potentially involving *experts*.

1   *Specify Search*. At the outset of MA, a *Scoping Review* is carried out to check the size, scope, and range of knowledge either by mapping key concepts across the literature or in *consultation with experts*. To delineate the topic, searchable terms suitable for systematic review are specified, being neither too narrow (of limited use) nor too diffuse (hence not feasible). Results can be documented in the top layer of a PRISMA diagram. In addition, the methodology for all other stages of the MA includes that the statistical methodology should be pre-specified. This ensures that at each stage, appropriate information is collated to feed into later stages, and guards against *ad hoc* practices and other forms of researcher bias.

2   *Search Adequacy*. A *Systematic Review* is conducted to appraise comprehensively all relevant literature using explicit search criteria (specified in Stage 1), to assess the inclusion of each study (Gough *et al.*, 2012). Following AMSTAR2 criteria helps to avoid *ad hoc* modification of search criteria. This makes it possible to quantify prevalence of study characteristics. Often, it is necessary to consult only the abstract of each study. In addition to traditional bibliographic searches of research literature, there are automated search engines, such as Twitter scrapers, global repositories, or big data streams. Furthermore, some studies might seed snowball sampling to increase coverage not reached by using keywords. Consultation with experts also might lead to identifying relevant studies. Transparent specification of a purely systematic or compound search strategy can be depicted in a PRISMA diagram.

3   *Study Eligibility*. Often, careful reading of each study is required to ascertain whether it triggers exclusion criteria. Novel MA might require more intensive effort, especially the construction of a conceptual framework to embrace emerging topics, and might benefit from expert consultation. A *Narrative Review* is done to summarise existing knowledge, define concepts, and identify relevant theory, whereas using a *Qualitative Review* requires critical analysis to synthesise such information (Grant and Booth, 2009). *Model-Centric*

Literature Review is a narrative review organised to align with an over-arching, statistical model (Jahandideh *et al.*, 2018).

4   *Consistent Quantification.* This relies on harmonisation of quantities across studies and characterisation of study quality. Such quantities may include effect sizes, group averages, counts, or totals. The challenge is that each study might report quantities using different conventions, which are not comparable until rescaled or transformed. Moreover, for a MA the option may be taken to allocate equal or unequal weight "democratically" to every study. Using both AMSTAR2 criteria and expert consultation methods provides consistent, repeatable, and transparent ways to achieve harmonisation and weighting (Fisher *et al.*, 2015).

5   *Statistical MA.* The culmination of data collation and preparation in the previous four phases is to apply a statistical model to pool, average, sum, or compare the harmonised quantities, accounting for varying study quality and sources of bias. Each study should represent an independent information source, and hence duplicates should be omitted. Typically, fixed or random effects models are fit for the purpose, using classical (frequentist) inference in software such as RevMan (The Cochrane Collaboration, 2014). Studies may be weighted depending on sample size, standard error, or some other measure of study accuracy (Lantz, 2013). Alternatively, *Bayesian Statistics* provides a natural framework for incorporating both published quantities as well as expert assessments (Low-Choy *et al.*, 2017). More complex statistical models can be used, such as structural equation models, integrative data analysis, or data fusion (Yu and Zeng, 2018).

6   *Interpretation.* This endpoint benefits from the use of a mixed-methods approach, accounting for qualitatively and quantitatively assessed shortcomings and risks of bias. The interpretation will also depend on the statistical paradigm: whether calculation of a pooled estimate to reduce prediction error (e.g. by using least squares or risk assessment); or the most plausible estimate (Bayesian); or an estimate that makes the data most likely (frequentist, maximum likelihood).

7   *Broader View and Implications.* From a general perspective, it is necessary to assess results in terms of what was omitted. For this reason, AMSTAR2 advocates an assessment of publication bias, e.g. using funnel plots to evaluate whether studies are relatively more similar than would be expected across independent studies. At this stage, the results of a *Realist Review* could provide a meta-narrative that enables reflection on policy and/or refinement of theory as well as an opportunity to reflect on the broader implications of MA results (Tricco *et al.*, 2011).

Research approach and design

MA can be viewed as a mixed method in a loosely coupled way, for instance, if review and MA are conducted separately (Creswell and Clark, 2011), or in a closely coupled way, if all phases are informed by both qualitative and quantitative research methods (Hesse-Biber, 2015, Chapter 1). To provide an expository view of the meta-analyses in this chapter, a case study (CS) methodology was adopted. Standard methods of MA, which rely on LR that is systematic, are prevalent in medicine and health. However, these methods are not so relevant to situations reflected in these case studies, where the use of MA was pioneering for the topic in each case. Case Study B (CSB) relies on multiple kinds of LR to source relevant studies, whereas in Case Study A (CSA) it was possible to curtail intensive LR by referring to a global repository of studies, refined by experts.

The two case studies illustrate how the same seven stages of MA can be put into practice, although they adopt quite different approaches, tools, and methodologies (summarised in Table 15.1). Yet, both perform similar steps (1. scoping, 2. ensuring adequate search, 3. checking eligibility) to collate and confirm study exclusion/inclusion. From an operational perspective, a major difference was that in CSA, a statistical methodology, mentioned earlier, available in a spreadsheet was used, whereas in CSB, an off-the-shelf approach, available in the open-sourced software RevMan was harnessed.

Results and discussion

Case Study A: Counting species living on coral reefs – Bayesian/risk assessment MA

Description. A published MA (Fisher *et al.*, 2015) estimated the total number of species inhabiting coral reefs worldwide, to inform environmental management and policy priorities. Previous estimates had relied on indirect analogy from a terrestrial rainforest or an aquarium. These estimates provided the first comprehensive baseline for evaluating change over time based on species counts from a marine taxonomy database. This non-standard MA used expert assessments to update and quantify uncertainty around these estimates, for aggregating rather than averaging counts across groups of marine taxonomy studies.

1   *Specify Search.* This MA relied on global database WoRMS, which has been acknowledged as a repository for all marine taxonomic findings for several decades, e.g. identification of new species and lumping/splitting existing taxa. Embedded into the professional practice of marine taxonomists, since 2007, WoRMS systematically records all known information sources on marine taxonomy. Hence WoRMS provided an ideal source (richness) for counting species in taxonomic groups.

2   *Search Adequacy.* Consultation with experts questioned the currency of WoRMS, highlighting that in some taxonomic groups, its information could be more incomplete and less accurate than others (e.g. fungi compared to fish), especially where species had lower commercial or recreational value, or lower public or scientific profile.

3   *Eligibility.* Experts were selected according to their expertise (Caley *et al.*, 2014; Fisher *et al.*, 2015). Since WoRMS embeds marine taxonomic standards,

Table 15.1 For two case studies: an overview and comparison of characteristics, including AMSTAR2 criteria (numbered), highlighting each methodology, for literature review or eliciting expert knowledge (*italics*)

Characteristic	Case Study A (Fisher et al., 2015): *How many species live on coral reefs?*	Case Study B (this chapter, 2020): *Is big data analytics (BDA) good for business?*
Field of meta-analysis	Ecosystem health, coral reefs	Information technology, systems
Status	Published meta-analysis	New meta-analysis of published studies
Purpose	Aggregate total species richness worldwide	Pool effect of BDA, across industry sectors/countries
Field per study	Marine taxonomy	Business/IT
1 Search strategy	Quality assured *systematic review* via global repository; research ethics for expert consultation	*Scoping and systematic review*, AMSTAR 2 criteria, and Review Manager Software (RevMan) workflow
2 Search adequacy	Marine species grouped by taxa, clustered by local expert	*Systematic review* ensures each study concerns big data analytics and business performance
3 Eligibility	Confirm taxon relevant to coral reef and expert	*Model-centric review* to ensure similar input and output variables used to analyse effect via structured equation model (SEM)
4 Quantities harmonised	*Experts* revise counts recorded in global database	Standardised effects
5 Uncertainty represented	*Bayesian approach* to elicit expert knowledge	Sample size, standard error (reported or inferred)
6 Statistical method to pool	*Risk assessment* method of linear pooling in spreadsheet	Random effects model in RevMan
7 Interpretation	Adopt *Bayesian methods* to report plausible total species count with error; compare with earlier estimates using *forest plot*	Test differences amongst studies via *hypothesis testing*, find consistent effect with *confidence intervals*
8 Implications	Quantify value of coral reefs to biodiversity, including *realist review*	*Funnel plot* used to assess publication bias
Mixed methods	Expert consultation (qual & quant) supplemented quant systematic review, and formed the basis for quant MA	Qual phase of systematic review followed by quant/ qual harmonisation of quant (SEM) studies, for input into quant MA

Source: Original.

each species recorded in the database is, by definition, eligible to be counted. It was necessary to remind taxonomists to focus solely on taxa (within their area of expertise) and estimates of species richness related to species that use coral reefs.

4   *Harmonising Quantities*. Experts were interviewed to capture their considered judgements on how results obtained from WoRMS ought to be adjusted. Applying expert elicitation and encoding methods popular in Bayesian statistics (O'Leary *et al.*, 2015), experts were asked to perform uncertainty analysis on species counts for each taxonomic group within their expertise. This protocol was embedded into a spreadsheet to enable real-time, *in situ* feedback (Fisher *et al.*, 2015).

5   *Uncertainty Represented*. Using the WoRMS estimate as a benchmark, experts were also asked for lower and upper bounds. Experts were guided to consider taxonomic coverage for three situations: (K)nown species named and databased in WoRMs, (D)iscovered but not yet named (i.e. pending databasing or lumping/splitting), or (U)ndiscovered (e.g. due to geographic or taxon-specific limitations of sampling). Assessments of each component were based on their career-long expertise in that taxon. Expert uncertainty was captured at face value, regarding plausible bounds and applying well-established risk assessment. The expert elicitation methods, which were piloted with ethics approval, ensured reproducibility across tens of marine taxonomists (O'Leary *et al.*, 2015).

6   *Statistical Methodology*. The elicitation protocol was proposed to feed into a risk assessment method for aggregating expert estimates using linear pooling (O'Leary *et al.*, 2015), and made reproducible using a published spreadsheet (Fisher *et al.*, 2015).

7   *Interpretation*. The final estimate was that 830,000 species use coral reefs (Fisher *et al.*, 2015), and found a 95% chance that this value actually lies between 550,000 and 1,330,000 (after accounting for uncertainty in data collection as well as model specification, according to the Bayesian approach). A forest plot was used to compare species richness and uncertainty across taxonomic groups, showing that species-rich taxonomic groups using coral reefs had varying uncertainty, where round worms had the highest species richness and uncertainty; whilst sea slaters and mollusca were the next richest taxa but much more well studied.

8   *Implications*. Since novel methodology was used, publications were reported separately on statistical (O'Leary *et al.*, 2015) and ecological aspects (Fisher *et al.*, 2015). Discussions of implications were addressed separately from the ecological viewpoint, and in terms of relying on expert knowledge. Risks of bias owing to study omission were addressed using the uncertainty model (Stage 4), which asked experts explicitly to consider for each taxon how many species remained potentially undiscovered.

9   *Mixed Methods*. Most research team members had some expertise in qualitative and quantitative research methods. Additionally, some researchers had traditional biological background. Interestingly, given the novel methodology

required at this interface of qualitative/quantitative methods, all publications reflected a mixed-methods approach, though emphasising the quantitative elements, where elicited expert knowledge was quantified for input into the quantitative models.

Case Study B: Big data and firm performance

Big Data refers to the enormous set of data characterised by large volume, extensive variety of data types, and the enormous speed with which these data are generated (Almeida, 2018). According to Sun *et al.* (2018), from a business perspective, Big Data offers a great opportunity for companies to extract effective business intelligence from this data. The aim of this CS was to combine LR and MA to identify the relevance of factors that affect the adoption of big data analytics (BDA) in an organisation and explore the dimensions of their impact on organisational performance. For this purpose, and to make it possible to carry out a systematic analysis of the importance of these factors, only quantitative studies that adopted a structural equations model were considered as a representation and data exploration model. Figure 15.1 shows schematic presentation of the PRISMA Flow Diagram. The data collection process covered publications indexed only until the end of October 2019.

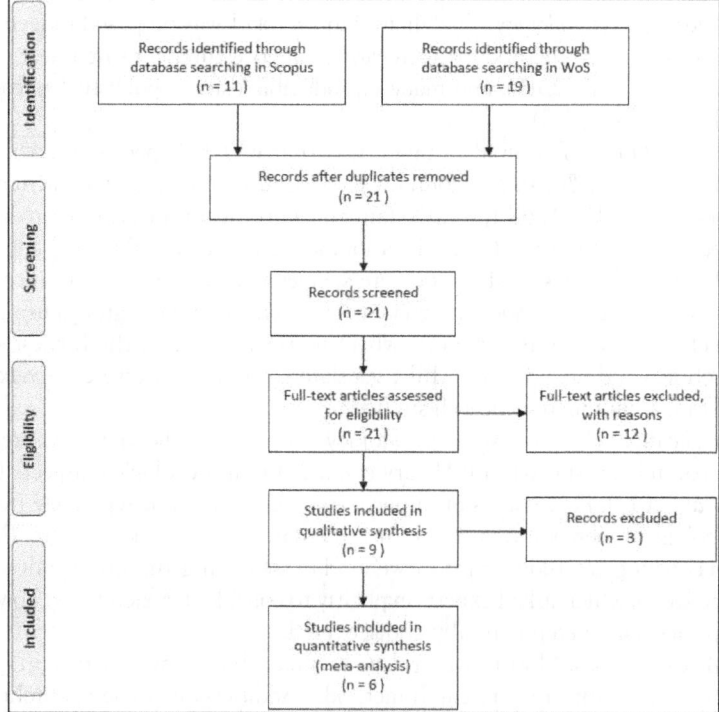

Figure 15.1 PRISMA flow diagram for systematic literature review in Case Study B.

1   and 2. *Study Selection using Scoping and Systematic Review.* Initially, a bibliographic search was conducted on Scopus and Web of Science databases using compound search terms "big data" and "firm performance". This was refined to focus on a homogeneous subset of quantitative studies by searching "structural equation model". The PRISMA diagram (Figure 15.1) illustrates how 21 records were obtained after duplicates were removed. The first study published in this area appeared only in 2017, which reveals the recent emergence of this area of knowledge.

3   *Eligibility using Qualitative Review.* In the process of assessing eligibility of studies, it became necessary to conduct a full-text article assessment. Careful reading was required to interpret the input and output variables and relationships across studies. Based on the assessment, nine studies were identified that were focused specifically on the role of big data technologies in organisational performance. In contrast, 12 studies were excluded because other factors were included, such as human resources capabilities, BDA management, knowledge management orientation, and supply chain performance, among others. Subsequently, three more articles were excluded because they included exploration of the role of big data technologies in complementary dimensions of organisational performance, firm adaptability, and environmental sustainability. In the end, for the MA presented in this chapter, six studies were included, as described briefly in Table 15.2.

4   *Consistent Quantification.* Standardised coefficients were extracted for the direct effect of BDA on firm performance. (Indirect effects and effects of other predictors will be examined in future work.) Corresponding standard errors were extracted also, or inferred from the published sample size, as well as *t*-statistic or *p*-value.

Table 15.2 Description of studies included in meta-analysis

Study	Description
Anwar *et al.* (2018): 312 Chinese firms	Relates Big Data technological capabilities and personal capabilities to competitive advantage and firm performance
Ferraris *et al.* (2019): 88 Italian SMEs	Relates Big Data analytics capabilities, knowledge management orientation, and firm performance
Gupta *et al.* (2019): 231 employees in companies of varying size	Relates Big Data predictive analytics, cloud ERP, and firm performance
Irfan and Wang. (2019): 240 firms in F&B industry, Pakistan	Relates impact of flexible IT resources, data assimilation, and internal/external integration on competitive performance
Mikalef *et al.* (2020): 202 IT managers/CIOs in Norwegian companies	Relates Big Data analytics capability, dynamic capabilities, marketing capabilities, technological capabilities, and competitive performance
Shan *et al.* (2019): 219 employees in diversified industries	Relates IT technology resources and capabilities, IT relationship resources, idle resources, strategic flexibility, compatibility, and competitive advantage

Source: Original.

5   *Quantitative Results.* The Review Manager v5.3 Software (The Cochrane Collaboration, 2014) estimated a pooled (direct) effect of BDA on firm performance at 0.34, with bounds (95% CI of 0.20–0.47) just inside lower estimates of three Asian studies (under 0.20) and higher estimates of European studies and one Asian study (0.3–0.5). The test for overall effect showed strong evidence against a zero overall effect ($Z = 4.85$, $p < 10^{-5}$), and standard deviation between studies was relatively small (Tau = 0.1). Given the small number of studies, the MA performed adequately ($Chi^2 = 9.52$ on 5 df) but exhibited some heterogeneity ($I^2 = 0.47$).

6   *Study Evaluation.* The AMSTAR2 guidelines were used to assess the validity and reliability of the MA with systematic LR (results not shown). Several elements were assessed, including inclusion and exclusion criteria, response rate, sampling method, and risk of bias, among others. Only the Gupta *et al.* (2019) study did not present information regarding the initial survey sample size and response rate.

7   *Implications.* Finally, the funnel plot (Figure 15.2) showed that these studies were more consistent than expected, although larger effect sizes co-occurred with slightly lower uncertainty. No studies found negative or negligible effect sizes. This was not unusual since it is difficult to publish structured equation models (SEMs0 with non-significant relationships).

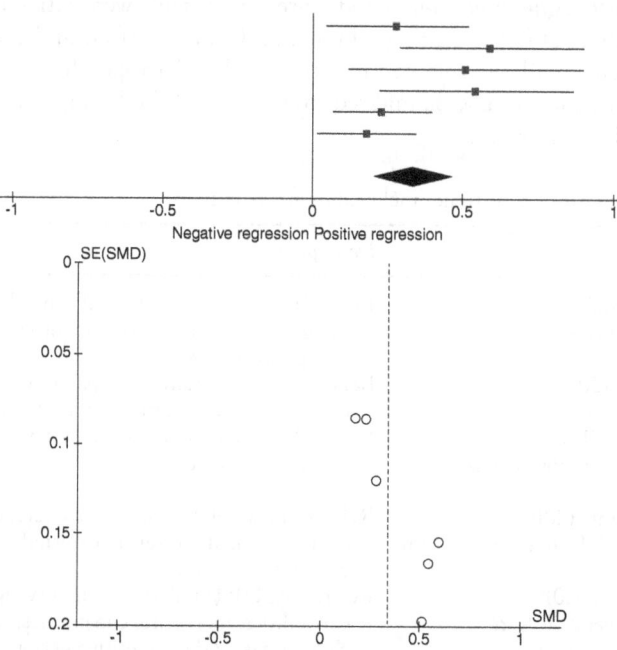

Figure 15.2 Top: Forest plot showing standardised effect sizes (x-axis) for each study (red dots) and confidence intervals (lines) for individual studies (y-axis); pooled estimate across studies (diamond). Bottom: Funnel plot showing standardised mean difference (SMD) of each study (x-axis) against its standard error (y-axis).

Implications for practice and research

Benefits of viewing meta-analysis as mixing methods

The case studies presented earlier illustrate the potential roles of qualitative and quantitative methods in MA. They show that mixing methods require an understanding of how to combine each qualitative and quantitative component purposefully (Rose and Low-Choy, 2019). In turn, these roles guide research design, pertinent specifically to LR, statistical analysis, and the interfaces between them. Moreover, substantial gains can be made by treating each component of MA as a mixed method rather than treating each component as distinctly qualitative or quantitative, as in a "Creswellian" mixed-methods research design, as defined in Creswell and Clark (2011). Instead, a broader perspective of mixed and multi-method can be useful for MA. A few gains have been highlighted below (for more in-depth discussion of these points or others, see Hesse-Biber and Johnson, 2015).

Statistical paradigm: computation and inference

For most MAs, a classical random effects analysis is adopted to pool estimates across studies. When this is the case, the accuracy of the estimated variance component across studies is typically estimated via resampling, a type of sensitivity analysis. An alternative is to use Bayesian statistics. However, this is not yet embedded into tools such as RevMan, making it less accessible. Fundamentally, Bayesian inference is defined as a formal mechanism for updating knowledge. This is often well suited to the purposes of MA, where the purpose is to collate the current state of knowledge (Low-Choy *et al.*, 2017, Section 8). In particular, as shown in CSB, expert assessments can be incorporated readily into a MA, following procedures of expert elicitation from Bayesian statistics that quantify and represent uncertainty statistically (e.g. Low-Choy *et al.*, 2009).

Meta-analysis of structural equation models

Standard criteria for appraising MA have been constrained historically to consider studies with certain statistical designs and/or analyses. For instance, the Cochrane Criteria and original AMSTAR criteria were tailored to apply to studies that implemented randomised control trials, case control designs, or other specific kinds of non-randomised designs. Neither CS fulfilled these constraints: CSA supplemented databased studies with spoken discussion of elicitation/risk assessment; and CSB focused on studies that used SEM rather than regression to analyse effects. This has benefits and drawbacks. On the positive side, it is useful to meta-analyse SEMs, since the complex relationships modelled using SEMs are communicated easily through diagrams, which also support analysis. Importantly, SEMs force researchers to be explicit about assumptions regarding causality or association. On the negative side, since SEMs can be more complex in structure than regressions, this might complicate MA (not shown here).

Appraising meta-analysis

Through association with the Cochrane Centre, AMSTAR2 criteria for MA are popular in many fields, including software engineering. Provision of the AMSTAR2 criteria online encourages meta-authors to check that their MA satisfies minimal best practice criteria. PROSPERO (Prospective Register of Systematic Review) is an online repository that can be used to record systematic reviews (Farrah *et al.*, 2019).

Uptake

From a practical perspective, the uptake of any research method relies on availability of easy-to-use tools that ideally provide a guided and managed workflow for implementing the method. RevMan (The Cochrane Collaboration, 2014) and other software packages (e.g. Meta-DiSc, Stata, SAS, R) enable easy collation of data and use of a classical, statistical MA. Although the spreadsheet used for the original Coral Reef CS to record, collate, and analyse the data used in the MA was published, it relied on a global database to compile and manage the individual studies. This presents an additional conceptual and practical hurdle to researchers wishing to apply a similar approach in different fields.

Conclusions

The AMSTAR criteria for appraising MA have recently been generalised to consider a wider array of study designs, beyond those commonly used in the medical or health arena. This is timely, and greatly widens the potential for more pioneering types of MA that include studies which are observational rather than experimental with randomised or pseudo-randomised designs. Some built environment disciplines (e.g. construction management and property management) are based on observational data and, whilst MA of this nature is rare in the built environment, the two cases presented in this chapter serve as good examples of how MA studies can be applied across different built environment disciplines. This provides an opportunity for selecting and extending the statistical impact of the many studies that have been conducted across these built environment disciplines. Indeed, in this chapter it has been demonstrated that the several domains of the revised AMSTAR2 criteria are relevant to describing non-standard meta-analyses.

A key message is that, to facilitate MA, LRs need not be systematic. Instead, multiple forms of LR methodologies (scoping, systematic, narrative, qualitative, model-centric, realist as well as expert assessments) can assist in several phases of MA, from scoping and collating studies, then assessing study eligibility, to implications.

The benefits of viewing MA as a mixed method were explored in this study. For instance, quantitative methods, such as expert elicitation of quantities, might help to sharpen several phases of LR. Equally, qualitative methods are fundamental to ostensibly quantitative stages, such as harmonisation of quantities prior to MA. Thus, a mixed-methods approach can integrate quantitative and qualitative aspects of MA more meaningfully.

References

Almeida, F. (2018). Big data: Concept, potentialities and vulnerabilities. *Emerging Science Journal*, 2(1), pp. 1–10.

Anwar, M., Khan, S.Z. and Shah, S.Z.A. (2018). Big data capabilities and firm's performance: A mediating role of competitive advantages. *Journal of Information & Knowledge Management*, 17(4), p. 1850045.

Caley, M.J., Fisher, R. and Mengersen, K. (2014). Global species richness estimates have not converged. *Trends in Ecology & Evolution*, 29(4), pp. 187–188.Castro, F.G., Kellison, J.G., Boyd, S.J. and Kopak, A. (2010). A methodology for conducting integrative mixed methods research and data analyses. *Journal of Mixed Methods Research*, 4(4), pp. 342–360.

Cheek, J. (2015). It depends: Possible impacts of moving the field of mixed methods research toward best practice guidelines. In: Hesse-Biber and Johnson (eds.), *Designing and Conducting Mixed Methods Research*, pp. 624–636, Oxford Handbooks, Oxford, UK.

Creswell, J. and Clark, V.L. (2011). *Designing and Conducting Mixed Methods Research*, Sage Publications, Thousand Oaks, CA.

Cooper, H., Hedges, L.V. and Valentine, J.C. (2019). *The Handbook of Research Synthesis and Meta-analysis*, Russell Sage Foundation, New York.

Farrah, K., Young, K., Tunis, M.C. and Zhao, L. (2019). Risk of bias tools in systematic reviews of health interventions: An analysis of PROSPERO-registered protocols. *Systematic Reviews*, 8(1), p. 280.

Ferraris, A., Mazzoleni, A., Devalle, A. and Couturier, J. (2018). Big data analytics capabilities and knowledge management: Impact on firm performance. *Management Decision*, 57(8), pp. 1923–1936.

Fisher, R., O'Leary, R.A., Low-Choy, S., Mengersen, K., Knowlton, N., Brainard, R.E. and Caley, M.J. (2015). Species richness on coral reefs and the pursuit of convergent global estimates. *Current Biology*, 25(4), pp. 500–505.

Gough, D., Thomas, J. and Oliver, S. (2012). Clarifying differences between review designs and methods. *Systematic Reviews*, 1(1), p. 28.

Grant, M.J. and Booth, A. (2009). A typology of reviews: An analysis of 14 review types and associated methodologies. *Health Information & Libraries Journal*, 26(2), pp. 91–108.

Gupta, S., Qian, X., Bhushan, B. and Luo, Z. (2018). Role of cloud ERP and big data on firm performance: A dynamic capability view theory perspective. *Management Decision*, 57(8), pp. 1857–1882.

Hesse-Biber, S. (2015). Mixed methods research: The "thing-ness" problem. *Qualitative Health Research*, 25(6), pp. 775–788.

Hesse-Biber, S. and Johnson, R. (2015). *Oxford handbook of multimethod and mixed methods research inquiry*. Abby Gross, Oxford, UK. Irfan, M. and Wang, M. (2019). Data-driven capabilities, supply chain integration and competitive performance. *British Food Journal*, 121(11), pp. 2708–2729.

Jahandideh, S., Kendall, E., Low-Choy, S., Donald, K. and Jayasinghe, R. (2018). The process of patient engagement in cardiac rehabilitation: A model-centric systematic review. *Behaviour Change*, 35(4), pp. 185–202.

Lantz, B. (2013). The large sample size fallacy. *Scandinavian Journal of Caring Sciences*, 27(2), pp. 487–492.

Low-Choy, S., O'Leary, R. and Mengersen, K. (2009). Elicitation by design in ecology: Using expert opinion to inform priors for Bayesian statistical models. *Ecology*, 90(1), pp. 265–277.Low-Choy, S., Riley, T. and Alston-Knox, C. (2017). Using Bayesian statistical modelling as a bridge between quantitative and qualitative analyses: Illustrated via analysis of an online teaching tool. *Educational Media International*, 54(4), pp. 317–359.

Mikalef, P., Krogstie, J., Pappas, I.O. and Pavlou, P. (2020). Exploring the relationship between big data analytics capability and competitive performance: The mediating roles of dynamic and operational capabilities. *Information & Management*, 57(2), pp. 1–15.

O'Leary, R.A., Low-Choy, S., Fisher, R., Mengersen, K. and Caley, M.J. (2015). Characterising uncertainty in expert assessments: Encoding heavily skewed judgements. *PloS ONE*, 10(10), p. e0141697.

Paré, G., Trudel, M.C., Jaana, M. and Kitsiou, S. (2015). Synthesizing information systems knowledge: A typology of literature reviews. *Information & Management*, 52(2), pp. 183–199.

Rose, J. and Low-Choy, S. (2019). Modern pedagogical approaches to teaching mixed methods to social science researchers. *HEAD'19. 5th International Conference on Higher Education Advances*, Valencia, Spain, pp. 1375–1382.

Shan, S., Luo, Y., Zhou, Y. and Wei, Y. (2019). Big data analysis adaptation and enterprises. Competitive advantages: The perspective of dynamic capability and resource-based theories. *Technology Analysis & Strategic Management*, 31(4), pp. 406–420.

Shea, B.J., Reeves, B.C., Wells, G., Thuku, M., Hamel, C. and Henry, D.A. (2017). AMSTAR 2: A critical appraisal tool for systematic reviews that include randomised or non-randomised studies of healthcare interventions, or both. *BMJ*, 358, p. j4008.

Sun, Z., Sun, I. and Strang, K.D. (2018). Big data analytics services for enhancing business intelligence. *Journal of Computer Information Systems*, 58(2), pp. 162–169.

The Cochrane Collaboration. (2014). *Review Manager (RevMan) 5.3 user guide*, The Nordic Cochrane Centre, Copenhagen. Available from: https://dplp.cochrane.org/sites/dplp.cochrane.org/files/public/uploads/rm5userguide.pdf.

Tricco, A.C., Tetzlaff, J. and Moher, D. (2011). The art and science of knowledge synthesis. *Journal of Clinical Epidemiology*, 64(1), pp. 11–20.

Yu, X.T. and Zeng, T. (2018). Integrative analysis of omics big data. In T. Huang (ed.), *Computational Systems Biology: Methods in Molecular Biology*, vol. 1754, Humana Press, New York, pp. 109–1355.

16 Analysing secondary data to understand the socio-technical complexities of construction-design decision-making

Payam Pirzadeh, Helen Lingard, and Nick Blismas

Introduction

Researchers in construction have identified considerable benefits of integrating construction expertise and knowledge into early project decision-making (Lingard *et al.*, 2014; Song *et al.*, 2009). Improved constructability and health and safety (H&S) have been highlighted frequently among other benefits. Early-stage collaboration and effective interaction within and between design and construction participants are vital to making construction process knowledge accessible to design decision-makers. Nevertheless, collaboration and effective interaction still seem to be a problem in practice and, in many cases, efforts to promote collaborative interactions fail to address the complex and dynamic nature of the design process.

Exploring the interactions between project participants while making decisions helps to understand the way in which construction knowledge is used during the design process. This understanding reveals the particular features of the interactions which support collaborative decision-making, leading to improved constructability and H&S outcomes. The patterns of interactions and information exchanges between project participants can be conceptualised as social networks. According to Pryke and Smyth (2006), through these networks, individuals involved in different project functions (e.g. planning, design, construction) exchange information, co-ordinate their tasks and establish a sense of mutual understanding about terminology, values, and priorities. Austin *et al.* (2007) suggested that collaborative design needs an easy flow of information between participants outside the rigid structures imposed by contractual arrangements. Participants' engagement in informal interactions creates the flexibility needed to adjust knowledge transactions to the particular needs of specific design decisions. Consequently, better informed decisions can be made.

For the study provided in this chapter, a social network perspective on design decision-making was adopted. The study drew upon data from a previous research project reported by Lingard *et al.* (2014). A comprehensive dataset collected during the previous research project was re-analysed. The dataset included 23 case studies, each focused on the design process of a construction project element, e.g. steel structure. The original research explored the overall interactions and information exchanges during the design of these elements. The aim was to understand what characteristics of the interactions were linked to positive H&S outcomes. However, the nature of this link was not explored in detail, because the analysis took a static view of interactions by aggregating the interactions over the whole design process and using descriptive network measures to understand patterns of interaction.

In contrast, the new study built on this previous research and extended it by providing detailed evidence, linking specific interaction features to decision outcomes. The aim of the new study was to unpack the mechanisms by which positive influence on decision outcomes occurs through intra-team communication in construction projects. To achieve this, it was necessary to examine systematically each decision circumstance, the knowledge inputs, and the way in which solutions were devised, to understand the effects on constructability and H&S outcomes.

Consequently, a new approach was used to enable a more refined analysis of the data. For the new study the qualitative and quantitative data were combined and a multi-level perspective was adopted to investigate the interdependence between social interactions and decision outcomes at different levels. The multi-level conceptualisation made it possible to capture and recognise three types of interdependence in the design process as follows (illustrated in Figure 16.1):

- At the macro-level, design decisions and the interdependence between them form a technical network.
- At the micro-level, design participants and the information exchanges between them create a social network.
- At the meso-level, participants' involvement in (and influence on) design decisions forms a two-mode, socio-technical network.

Furthermore, interaction patterns were studied at each decision point to identify changes during the decision-making process. Thus, by combining multi-level and longitudinal perspectives during the new study, it was possible to capture and understand the complexities and dynamics of building-design decision-making

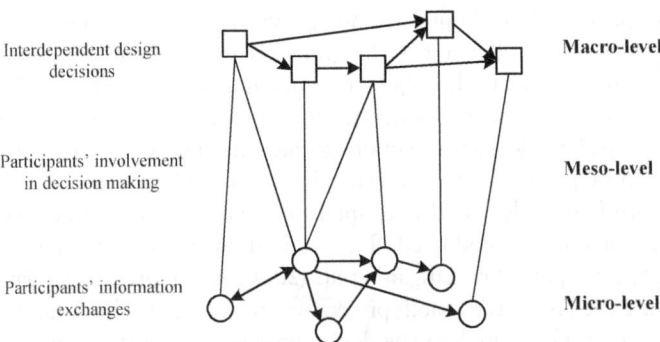

Interdependent design decisions   Macro-level

Participants' involvement in decision making   Meso-level

Participants' information exchanges   Micro-level

Figure 16.1 A multi-level network capturing interdependent design decisions and their associated interactions.

and the social interactions underpinning it. During the secondary analysis of the data, a statistical network technique was used in combination with qualitative analytical techniques. This made it possible to investigate the "building blocks" of social interaction networks. This understanding helped to explain how and why social interactions influence decision outcomes.

When using secondary data, it is important to ensure that the existing data is suitable for answering the new research questions. In the new study, this was achieved by: (1) ensuring an alignment between the objectives of the new study and the original study, for which the dataset had been collected; (2) using a case study approach in the new study similar to the original study; (3) applying case selection criteria in the new study that were similar to the original study; (4) considering the data types and data collection methods in the original study when developing the research design for the new study; and (5) ensuring high familiarity of the researchers with the original study and dataset as a result of their involvement in the original study. These considerations have been explained further in the section about research approach.

Secondary analysis can be conducted using a sub-set of the original dataset. This requires access to large and rich datasets as well as a clear set of selection criteria. In the new study, two additional case selection criteria were established (see the section on research approach) and applied to select a sub-set of six cases from the original dataset of 23 cases. The selected six cases were deemed to be most applicable to answering the research question in the new study. These criteria were used further to ensure the relevance and richness of data for each case and the context variability of the cases in the new study. This, in turn, contributed to the external validity and generalisability of the study. Figure 16.2 illustrates the case selection process and secondary analysis in the new study.

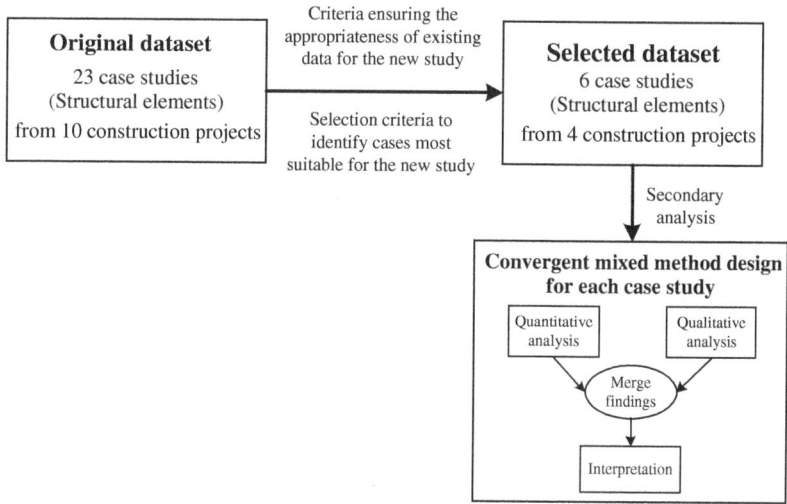

Figure 16.2 Case selection process and secondary analysis in the new study.

In the following sections, the complexities of construction-design decision-making are explained, followed by a description of the existing data and the approach used for its re-analysis. Then, a case study is presented illustrating the application of the analysis together with a discussion of the results and their implications.

The socio-technical process of design decision-making in construction

Construction-design decision-making is a complex process. The design complexities can be conceptualised at different levels. At a macro-level, inter-related decision-making episodes form a complex network. As design progresses, design decisions build on each other and new decisions require information from the previous ones. Lingard *et al.* (2011) indicated how design outcomes emerge from a network of inter-related decisions made through repeated interactions between multiple stakeholders. The research by Lingard *et al.* (2011) reflected also the complex structures of information exchanges that support the design decision-making process as well as the structures of governance that influence the roles, authority, and power relations between project participants. Similarly, Austin *et al.* (2002) illustrated the schematic design process in a construction project as a network of tasks linked by information flows. Austin *et al.* (2002) reported a network comprising some 150 tasks and 1,500 information flows. They further identified approximately 580 tasks and 4,600 information requirements in the detailed design process.

At a micro-level, complex social interactions underpin design decision-making. In a recent longitudinal case study, Pirzadeh and Lingard (2017) combined social network analysis and in-depth interviews to account for the dynamism and temporal nature of the building-design process and its underpinning interactions. As the study revealed, each decision-making scenario involved specific knowledge sources and interactions. Consequently, the participants in each decision, and the patterns of interaction between them, were specific to that scenario, with changes observed during the decision-making.

The social interactions take place within the context of decision-making and are inseparable from decision-making. Through the engagement of project participants in different decision-making scenarios and the social interactions between them, participants' knowledge is activated and made accessible to support decision-making. At the same time, the social negotiations among design participants facilitate accommodation of their diverse and transforming interests while design outcomes evolve (Tryggestad *et al.*, 2010).

Consequently, to gain a comprehensive understanding of the design process, it was necessary to investigate decision interdependence and interaction patterns simultaneously. A multi-level network perspective can facilitate this. The multi-level network approach is suitable for a number of reasons:

- It enables the simultaneous consideration of the interdependence between decisions and the social interactions between decision-making participants in one model.

- It facilitates consideration of complexities and relationships between social variables (interactions) and technical outcomes (decisions) that exist at different levels of analysis.
- It allows for the establishment of an association between social and technical aspects of the design process in construction project environments.

Research approach

A multiple case study approach was adopted for this study. Six cases were purposefully selected from an existing dataset of 23 cases. Each of these cases represented the building-design process of a structural element. A convergent, mixed-method design was used to re-analyse each selected case. This design involved merging the results of secondary, quantitative, and qualitative analyses during each case study to understand the new research problem from multiple perspectives (Creswell, 2015). To illustrate this approach, one case study is presented later in this chapter.

Applying a quantitative approach made it possible to investigate specifically the relationship between different features of intra-project, social interactions, and H&S-related technical outcomes in each case. In particular, from the analysis, different network features and configurations were identified and which configurations were linked to better H&S decision outcomes were determined. Social network analysis, statistical analysis, and data visualisation techniques were used to analyse the quantitative data. A quantitative approach has been used in previous research to study the relationship between complex, intra-project interactions and different aspects of project management, such as the effectiveness of contractual and performance incentives (Pryke, 2004); knowledge management (Brookes *et al.*, 2006); project team performance (Chinowsky *et al.*, 2008); project effectiveness (Chinowsky *et al.*, 2011); and level of profit performance (Park *et al.*, 2010). Recently, Lingard *et al.* (2014) used quantitative techniques to link the patterns of constructors' communication with other stakeholders, H&S performance.

During this study, further investigation was undertaken to understand how and why various interaction network configurations affect H&S-related decision outcomes. A qualitative approach was deemed to be suitable to answer the "how" and "why" questions. This approach led to a more complete understanding of the various views and explanations held by the project participants about design decisions and H&S, and helped to find out how project participants communicated their views during design decision-making and interacted to negotiate different project design outcomes. Qualitative analysis was used to examine design decisions, the reasons for making them, their interdependence, and the H&S outcomes arising from them.

The original data

The existing dataset consisted of transcripts of 185 in-depth interviews with key participants from ten construction projects in Australia and New Zealand. In each project, specific building, structural elements with H&S challenges were selected. The total number of elements in the dataset was 23.

The purpose of the interviews was to explore key decisions about the design of permanent features of a building or structure in each project, the construction process for the building or structure, and the way that H&S hazards were controlled during the construction process.

In addition, for each project element, data relating to social interactions between participants in the design decision-making were gathered. These data were collected by conducting additional interviews with key project participants. In each case, the participants were asked to rate the frequency of their interactions with other participants when making each design decision. The rating was done using a 5-point Likert response format, ranging from 1 (occasionally) to 5 (daily). In addition, the participants were asked to describe the importance of communication in their own words. This made it possible to identify relevant communication activities relating directly to design decision-making as opposed to general administrative communication. Responses to these questions were reviewed and coded qualitatively into the following categories according to whether the communication:

1   was not important to the design decision-making;
2   was good to know but had little impact on the design decision-making;
3   was good to know and was somewhat important to the design decision-making;
4   was necessary to know to contribute to the design decision-making;
5   was highly important and design could not proceed without this communication.

The participants also nominated and rated other influential participants in each decision-making scenario. A participant's decision-making power was calculated by adding up the ratings received. The results were scaled to range from 0 to 5. All the interaction data were formatted and saved in Excel spreadsheets.

Ensuring the appropriateness of the existing dataset for the new study

The existing data were deemed to be suitable for this new study for several reasons:

- *Alignment of research objectives*: The objectives of this study were in line with, and represented a substantial extension of, the objectives from the previous research project. In the previous research, the effect of overall interactions and information exchanges during design decision-making on H&S in construction projects was explored. From the initial research project, evidence was provided that positive H&S outcomes were facilitated by integrating construction process knowledge into design decision-making. Subsequently, the aim of the current study was to investigate this relationship further through understanding the characteristics of effective communication about H&S-related design decision-making.
- *Data types and data collection methods*: The dataset included both quantitative and qualitative data. These data types were combined within each

case. This combination proved to be highly synergistic (de Weerd-Nederhof, 2001), making it possible to discover different kinds of relationships and explain them, using the richness that comes from qualitative, anecdotal data (Mintzberg, 1979). Moreover, the detailed nature of the quantitative data, which was collected at each decision point, made the data suitable for the detailed analysis of changes in interactions and decision-making circumstances, which was a purpose of the current study. The qualitative data helped to explain the findings further.

- *Case selection criteria*: The initial research project used a case study approach. The case study data were collected using purposeful sampling. Purposeful sampling involves verifying that each case meets certain criteria before it is accepted in the sample. For a case to be included, the following criteria were established:

  - It had to be drawn from a construction project – ideally a live project to ensure participants would be able to focus directly on, and recall, the decision-making process, the communication related to it, and the decision outcomes;
  - It had to involve access to the design and construction participants – using a live project facilitated easier access to potential participants, where resource commitments to the project had already been made;
  - It had to be taken from projects using a variety of contractual or organisational arrangements;
  - It had to be taken from projects ranging in cost, starting from $250,000 upwards;
  - There had to be variation in the types of works (i.e. commercial, public, residential, industrial); and
  - The project had to present particular H&S challenges for construction.

Selecting a sub-set of six cases from the original data

The above criteria were in line with the aim of the new study, rendering the data suitable for analysis. Subsequently, additional selection criteria were applied to select six cases from the 23 existing cases. Establishing these criteria made it possible to identify the cases most suitable for the new study. The additional criteria included:

- Ensuring context variability by selecting the cases in a way that they represented different projects, construction sectors, procurement approaches, and H&S challenges.
- Ensuring the richness of the data and existence of comprehensive qualitative and quantitative data for each case.

An overview of the selected cases for the new study is presented in Table 16.1. For illustration, the results of Case Study 4 are reported later in this chapter.

Table 16.1 Overview of the six cases selected for the new study

Case ID	Case/structural element	Project	Procurement	Sector
1	A high-rise façade structure	Forty-two storey residential complex	Design and build	Residential
2	Roof and wall cladding		Design and build	Industrial
3	Roof structure	Manufacturing facility	Design and build	Industrial
4	Foundation system and steel structure		Design and build	Industrial
5	Basement structure	Cemetery mausoleum	Traditional	Commercial
6	Rehabilitation of steel structure	Food processing plant reconstruction	Accelerated design and build	Industrial

Source: Pirzadeh(2018).

Analysis

Content analysis was conducted on the raw qualitative data to identify and contextualise the design decisions, and to reveal the rationale for the decisions together with their chronology and interdependence. The H&S implications of the decisions were identified also from the interview contents. Examples included design features requiring workers to work in a confined space, on the exterior of a high-rise building, and in unsupported excavated areas. Narratives were developed describing the key decision-making episodes in each case. In addition, network diagrams were developed showing the interdependence of the decisions. The patterns of these interdependent relationships were studied by applying social network theory. This involved transforming qualitative data about the interdependence of decisions into quantitative network data.

In addition, interaction networks were created to map the frequency of communication between participants at each decision-making point. The nodes in these networks indicated participants involved at each decision-making point, and the connections between them reflected direct information exchanges. These interaction patterns were investigated at each decision point. The networks were aggregated for each case to capture the overall interaction pattern during the design process.

Subsequently, a multi-level network was created to capture and analyse simultaneously the technical interdependence between design decisions and the social interactions between participants in each case. The macro-level network consisted of the design decisions and the technical interdependence between them. As described earlier, this network was the outcome of qualitative data analysis.

However, the micro-level network represented the social interactions that took place between the participants during the design process. This network was the outcome of the quantitative data analysis. To link the two networks, a meso-level network was developed indicating the involvement of participants in decisions based on their decision-making power, i.e. a connection between a participant and a decision was established when the participant was involved in making the decision and had sufficient power to influence the decision outcome. To analyse the multi-level network, exponential random graph models were used. This approach makes it possible to identify significant configurations that characterise these networks. The results were interpreted with reference to the interview data. To illustrate this, the findings from Case Study 4 (Table 16.1) are presented in the following section.

Illustrative case study

The case involved the design and construction of the foundation system and steel structure for the storage facility of a manufacturing plant. Interviews had been conducted originally with 14 project participants. The project was procured using a design and construct (D&C) approach. At the early stage of the project, the client engaged a consultant to review the design of the client's facilities in other locations to capture the best design features of existing facilities. Based on this review, a generic design was developed with a strong focus on operations and end-use features of the facility as well as health regulatory requirements. To maximise the usable area, a steel structure consisting of five rows of columns and three spine trusses was specified in the concept design. The generic design and project specifications were handed over to the constructor. The constructor suggested eliminating one row of columns and revised the layout of the remaining columns. Consequently, it was necessary for fewer columns to be manufactured and installed. In addition, the number of pad foundations to support the columns was reduced. The constructor also revised the foundation design to pad foundations without reinforcement. Using this design, it was not necessary for the workers to install rebars in excavated areas. These design revisions improved the construction workers' H&S significantly. However, late changes by the client created extra work during construction.

Results of the illustrative case study and discussion

Construction input into the design decision-making

As revealed in the analysis, end-use requirements and compliance with health regulations were the main concerns during the concept design. Nevertheless, the D&C procurement approach, together with early involvement of the constructor, and the client's willingness to leave the constructor's team in charge of the technical design and construction decisions, provided a favourable environment for including construction expertise in the decision-making process. The interaction network patterns at key design decision points showed a central position for the constructor. The interview data disclosed further that the constructor used this position to influence the design process and to align the design decisions with their construction expertise.

Thus, constructability and H&S considerations were integrated into the design decision-making from the time the constructor was involved. Some examples were decisions about the layout of columns and eliminating one row of columns, off-site manufacture of structural elements, attaching bearing plates to the columns for temporary support of roof trusses during installation, eliminating reinforcement for the foundations, and the decision to cast-in baseplate anchor bolts and use a jig to fix them in place from outside the excavated area. In addition, the construction process was planned in conjunction with the structural design process. For instance, the sequence of activities planned for safely erecting the structure kept crane movements within safe distance, and the structure was designed so that each section was self-supporting to reduce the amount of propping and temporary works.

All of these design decisions greatly reduced workers' exposure to risks associated with falling from height or being hit by falling or moving objects. Overall, the constructability and H&S considerations during the design process facilitated implementation of mostly technological H&S risk controls in the construction stage.

Knowledge inputs and communication during the design process

The client's consultant defined and communicated design requirements to design and construction participants early in the conceptual design stage. Reviewing end-use and operational features of similar production facilities helped in identifying and articulating these requirements. The client and the consultant remained involved in the design process by establishing a formal design-approval process and participating in regular fortnightly design review meetings. In addition, the client's technical consultant participated in the interactions underpinning most key design decisions. These arrangements ensured that the client's objectives and requirements were considered during the design process.

Frequent and two-way information exchanges between the designer and constructor were an important factor in integrating design and construction decision-making. The two-way interactions started as early as deciding the layout of columns and foundation system, and continued throughout the design process. Reciprocal and frequent interactions between the constructor and design engineer underpinned almost all key design decisions.

Early engagement of the sub-contractors in the design decision-making was also observed in this case. The steel erectors and concreters were involved in the information exchanges as soon as the arrangement of columns and pad foundations were decided. Moreover, the interaction networks underpinning the key design decisions indicated that the sub-contractors remained involved in the decision-making process and influenced the outcomes of almost one-third of the decisions. Their input generally assisted in optimising material usage, in deciding on the construction process when making design decisions, and in addressing design variations when construction was already underway. In particular, when there was a design change, the sub-contractors were informed by the constructor to ensure they were up to date (e.g. when variations were made to the layout of the columns and pads).

There was frequent interaction between the design engineer, constructor, and client's consultant when making key decisions. Based on the overall pattern of interaction, the most frequent information exchanges took place between these three key participants, with the constructor acting as a "broker" (providing the only point of contact between other participants) during the information exchanges. These participants were the decision-makers for the three main aspects of the project: the design, construction, and end-use (commissioning). Therefore, regular interaction between them ensured a co-operative decision-making approach. In addition, the design process involved a relatively centralised pattern of interaction between participants. The pattern reflected a central role for the constructor who interacted directly with all participants; thus, the constructor had access to different sources of expertise and information within the network and was able to control the information flow and engage different participants as required, using their expertise to improve different aspects of design and construction. Nevertheless, there were occasions when timely communication between the constructor and client was not achieved (e.g. during propping and temporary works). As the data indicated, this issue was mainly as a result of the constructor feeling confident of being able to design and install the temporary works; thus, the constructor did not see that it was necessary to involve other participants, while the client was concerned about the effect that earthquake loads could have on the design of temporary works and wanted to check the details of temporary works.

Dealing with design variations

Late design variations were made in response to the client's request for changes to the building layout and operational (end-use) requirements. Although the changes did not introduce new H&S risks, they led to more work and increased workers' exposure to existing hazards such as falling from height, being struck by objects and material, and over-exertion and bodily reaction. The client acknowledged that pressure of time to complete the project was a factor that led to commencement of the construction process before the design work had been completed, resulting in variations to the design during construction. Nevertheless, the design changes were managed collaboratively to find the best way to address the changes with minimal impact on the construction process. For example, when deciding to add a new section to the building, the client's team asked the constructor's opinion about the area and layout for the new packaging section. In addition, making decisions about adding a packaging area to the facility and extending the structure, relocating existing columns, adding columns at the end of building, and installing additional columns for the packaging area involved the highest number of participants in comparison with other decisions.

Involvement in decision-making and alignment between participants' expertise and decision dependence

Regarding the pattern of involvement in decision-making, detailed multi-level network analysis revealed two significant effects (Figure 16.3). There was a statistically significant tendency for participants to be involved directly in making sets of interdependent decisions (network effect indicated in Figure 16.3a). In addition, the analysis revealed a significantly low tendency for individuals to interact when involved in different but interdependent decisions (network effect indicated in Figure 16.3b). Consequently, the analysis indicated that the network effect shown in Figure 16.3a was much more prevalent in the multi-level network than the network effect shown in Figure 16.3b. These results suggested that where there was a dependency between decisions in this case study, the decisions were more likely to be made through direct involvement of a number of common participants rather than through interaction between only them. These results indicated a match between participants' expertise and decision dependence and, more importantly, the empowerment of participants to influence the decision outcomes directly when their skills were relevant. This facilitated an efficient and direct transfer of knowledge and expertise when making technically dependent decisions.

Direct involvement of participants with relevant knowledge increases opportunities for using their expertise and tacit knowledge during decision-making. This improves the quality of the decision-making by making it possible to understand the design rationale (i.e. the "why"), and enhancing inter-disciplinary conversations and collaborative problem-solving in multi-disciplinary teams (Dossick and Neff, 2011).

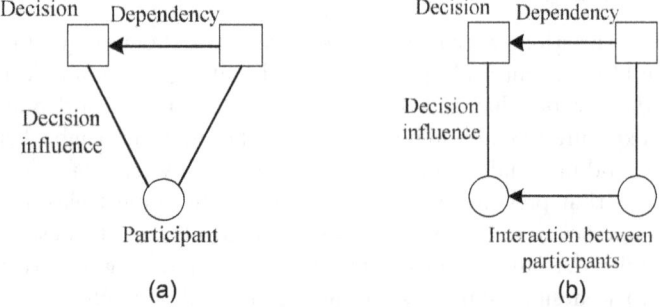

Figure 16.3 The significant multi-level network effects in the case study (squares represent decisions and circles indicate participants), (a) Tendency for participants to be involved in making interdependent decisions, (b) Tendency for participants involved in related decisions to exchange information without direct involvement in both of the decisions.

Implications for research and practice

In the secondary study reported in this chapter, an innovative multi-level network approach, which had not been used before in construction research to date, was applied to re-analyse existing data. Analysing both quantitative and qualitative data using longitudinal and multi-level network approaches made it possible to investigate the way in which network characteristics affect project outcomes (in this case, H&S outcomes). Using this approach, stronger relations were established between dynamic social processes and the project outcomes shaped by them. With this understanding, solutions can be identified to integrate constructability and H&S considerations into the design process effectively. As highlighted by Lingard *et al.* (2012), in the collective, reflexive, and uncertain context of design, it is imperative that any study undertaken to investigate the development of H&S-related design processes should take into consideration the reflexive and interactive nature of design work.

In this case study, the multi-level conceptualisation of design process facilitated an understanding of the socio-technical interdependence that characterises construction-design decision-making. Above all, research design provided strong evidence that positive H&S outcomes are realised when two-way interactions between participants in building design and construction coincide with the direct involvement and high influence of participants with construction expertise. The direct involvement of construction participants increases the opportunities for including their tacit construction knowledge (expertise), which is extremely hard to communicate otherwise.

In practice, understanding the characteristics of effective communication makes it possible to modify communication networks based on specific design and construction needs in each project. The evidence from the case study presented in this chapter suggests that communication activities during the design process need to be actively planned, monitored, and optimised to:

- engage sources of construction knowledge and expertise directly in decision-making and enable them to influence design outcomes actively;
- encourage collaborative information exchanges within sub-groups of participants at each decision-making episode; and
- facilitate two-way and direct information exchanges between participants and establish frequent communication between key influential participants for co-ordination of decision-making activities.

From a research perspective, the case study highlights how employing a new research design and adopting a novel multi-level network perspective made it possible to re-analyse an existing dataset in a more comprehensive way. The convergent, mixed-method design made it possible to combine qualitative and quantitative approaches. The multi-level network perspective, which had not been applied in construction research before, made it possible to link social interaction patterns to building-design decision outcomes explicitly and to explore different dimensions of construction-design decision-making, i.e. socio-technical complexity and dynamism. Consequently, it was possible to extend the original research significantly.

During the secondary study, comprehensiveness of the investigation was ensured by considering the design decisions, their circumstances, and their interdependence as well as by mapping and analysing the patterns of interaction that underpinned the construction-design decisions. The combined analysis and triangulation of rich qualitative and quantitative data enabled a unique, multi-faceted, and in-depth understanding of the socio-technical complexity and dynamism of construction-design decision-making and its associated H&S outcomes. Triangulation of the data also improved the reliability of the findings.

The suitability of using the existing data to answer the new research questions was another motivation for the second study. H&S-related design decision-making was the main topic of the interviews conducted in the original study. Likewise, the aim of the second study was to understand the characteristics of effective communication in the context of H&S-related design. Hence, the data were highly relevant to the secondary study. Nevertheless, it was still necessary to confirm the breadth, depth, completeness, and relevance of the dataset in the context of the new study (Sherif, 2018). This was done by establishing a set of criteria to ensure the appropriateness of the existing data for the second study, as explained in the methodology section of this chapter.

In addition, familiarity of the researchers with the parent study and the existing dataset was favourable in the context of secondary data analysis (Ruggiano and Perry, 2019). In this study, the researchers who undertook the secondary study were also involved in the original study and were highly familiar with the data collection approach as well as the research methodology and outcomes. This helped to recognise the limitations of the original study and to apply new theoretical perspectives to re-analyse the data to extend the original findings.

Conclusions

In the study presented in this chapter, a secondary research method was applied to re-analyse an existing, comprehensive dataset. The original dataset included 23 case studies. The appropriateness of the data for the secondary study was established by:

- understanding the context of the original study, considering the case selection criteria applied in the original study and the methods by which data had been collected; and
- ensuring that the original case selection criteria were also relevant in the context of the new study.

Furthermore, two additional selection criteria were used to select six cases from the original dataset, which were most appropriate to investigate the research question in the new study.

For the new study, a convergent, mixed-method design was adopted. Using this research design, the existing quantitative and qualitative data for each selected case were re-analysed and the findings were merged and interpreted. This approach made it possible to investigate multiple perspectives, cross-validate the findings, and obtain a more comprehensive understanding of the research problem in comparison with the original study.

In particular, a new multi-level network framework was applied to re-analyse the data and extend the findings of the original study. Thus, it was possible to investigate the socio-technical complexities of H&S-related design decision-making and to understand the characteristics of effective interaction, which provided knowledge support for construction-design decision-making and led to positive constructability and H&S outcomes. To illustrate this, the findings of one case study (out of the six selected cases) were reported in this chapter.

Understanding what makes communication effective in the context of design process can highlight opportunities for project management teams to improve the project communication activities. The results suggested that it is necessary for project teams to monitor and improve their interaction patterns continuously to encourage free-flowing information between participants and to ensure that participants with relevant construction knowledge and expertise are involved in decision-making directly and have the power to influence the decision outcomes.

References

Austin, S., Newton, A., Steele, J. and Waskett, P. (2002). Modelling and managing project complexity. *International Journal of Project Management*, 20(3), pp. 191–198.

Austin, S.A., Thorpe, A., Root, D., Thomson, D. and Hammond, J. (2007). Integrated collaborative design. *Journal of Engineering, Design and Technology*, 5(1), pp. 7–22.

Brookes, N.J., Morton, S.C., Dainty, A.R.J. and Burns, N.D. (2006). Social processes, patterns and practices and project knowledge management: A theoretical framework and an empirical investigation. *International Journal of Project Management*, 24(6), pp. 474–482.

Chinowsky, P., Diekmann, J. and Galotti, V. (2008). Social network model of construction. *Journal of Construction Engineering and Management*, 134(10), pp. 804–812.

Chinowsky, P., Taylor, J.E. and Di Marco, M. (2011). Project network interdependency alignment: New approach to assessing project effectiveness. *Journal of Management in Engineering*, 27(3), pp. 170–178.

Creswell, J.W. (2015). *A Concise Introduction to Mixed Methods Research*, Sage, Thousand Oaks, CA.

de Weerd-Nederhof, P.C. (2001). Qualitative case study research: The case of a PhD research project on organising and managing new product development systems. *Journal of Management Decision*, 39(7), pp. 513–538.

Dossick, C.S. and Neff, G. (2011). Messy talk and clean technology: communication, problem-solving and collaboration using building information modelling. *Engineering Project Organization Journal*, 1(2), pp. 83–93.

Lingard, H., Cooke, T. and Blismas, N. (2011). Who is 'the designer' in construction occupational health and safety? In: C. Egbu, and E.C.W. Lou (eds.), *Proceedings of the 27th Annual ARCOM Conference*, Association of Researchers in Construction Management, Bristol, pp. 299–308.

Lingard, H.C., Cooke, T. and Blismas, N. (2012). Designing for construction workers' occupational health and safety: A case study of socio-material complexity. *Construction Management and Economics*, 30(5), pp. 367–382.

Lingard, H., Pirzadeh, P., Blismas, N., Wakefield, R. and Kleiner, B. (2014). Exploring the link between early constructor involvement in project decision-making and the efficacy of health and safety risk control. *Construction Management and Economics*, 32(9), pp. 918–931.

Mintzberg, H. (1979). An emerging strategy of 'direct' research. *Administrative Science Quarterly*, 24(4), pp. 582–589.

Park, H., Han, S.H., Rojas, E.M., Son, J. and Jung, W. (2010). Social network analysis of collaborative ventures for overseas construction projects. *Journal of Construction Engineering and Management*, 137(5), pp. 344–355.

Pirzadeh, P. and Lingard, H. (2017). Understanding the dynamics of construction decision making and the impact on work health and safety. *Journal of Management in Engineering*, 33(5), p. 05017003.

Pirzadeh, P. (2018). *A social network perspective on design for construction safety*. Unpublished PhD thesis, RMIT University.

Pryke, S.D. (2004). Analysing construction project coalitions: Exploring the application of social network analysis. *Construction Management and Economics*, 22(8), pp. 787–797.

Pryke, S.D. and Smyth, H.J. (2006). *The Management of Complex Projects: A Relationship Approach*, Blackwell Publishing, Oxford.

Ruggiano, N. and Perry, T.E. (2019). Conducting secondary analysis of qualitative data: Should we, can we, and how? *Qualitative Social Work*, 18(1), pp. 81–97.

Sherif, V. (2018). Evaluating pre-existing qualitative research data for secondary analysis. *Forum: Qualitative Social Research*, 19(2). doi:10.17169/fqs-19.2.2821.

Song, L., Mohamed, Y. and AbouRizk, S.M. (2009). Early contractor involvement in design and its impact on construction schedule performance. *Journal of Management in Engineering*, 25(1), pp. 12–20.

Tryggestad, K., Georg, S. and Hernes, T. (2010). Constructing buildings and design ambitions. *Construction Management and Economics*, 28(6), pp. 695–705.

Appendix A: Chapter 14
Computation of cosine similarity for shared terms

Term (t)	$w_{A,t}$	$w_{C,t}$	$w_{A,t} \times w_{C,t}$	$w_{A,t}^2$	$w_{C,t}^2$
Energy	25.5150	8.4960	216.7755	651.0152	72.1821
Renewable	18.9020	4.5845	86.6563	357.2845	21.0178
Emissions	17.5746	8.2349	144.7255	308.8648	67.8143
Carbon	17.0672	13.8532	236.4352	291.2903	191.9103
New	16.7210	2.8943	48.3958	279.5908	8.3771
Solar	16.7210	2.8943	48.3958	279.5908	8.3771
Low	15.4379	2.2923	35.3877	238.3300	5.2544
Electricity	14.4494	4.4891	64.8654	208.7863	20.1523
Sustainable	12.6433	1.4472	18.2968	159.8520	2.0943
Change	11.7680	5.0706	59.6708	138.4863	25.7109
Fuel	11.7402	6.0211	70.6883	137.8316	36.2532
Power	11.7402	2.9101	34.1652	137.8316	8.4687
Sector	11.7402	2.9101	34.1652	137.8316	8.4687
Policy	10.4803	3.8801	40.6653	109.8375	15.0555
Potential	10.4803	3.3804	35.4277	109.8375	11.4271
Construction	9.6108	2.2923	22.0305	92.3678	5.2544
Consumption	9.0309	2.9101	26.2809	81.5572	8.4687
Demand	9.0309	1.4472	13.0691	81.5572	2.0943
Government	8.6497	3.4384	29.7411	74.8180	11.8225
Technology	8.6497	1.4472	12.5175	74.8180	2.0943
Study	8.1278	2.9101	23.6528	66.0613	8.4687
Environmental	7.7505	4.5845	35.5321	60.0698	21.0178
Costs	7.6887	2.9101	22.3748	59.1154	8.4687

(Continued)

Climate	7.2247	6.7337	48.6490	52.1966	45.3426
Increase	7.1962	2.2923	16.4955	51.7853	5.2544
High	6.7276	1.4472	9.7359	45.2603	2.0943
Research	6.6433	2.2923	15.2281	44.1329	5.2544
System	6.1682	6.7608	41.7017	38.0463	45.7082
Available	4.8165	1.4472	6.9702	23.1985	2.0943
Data	4.8165	1.4472	6.9702	23.1985	2.0943
Level	4.8165	3.4384	16.5609	23.1985	11.8225
National	3.0103	4.4891	13.5136	9.0619	20.1523
Average	1.8062	2.9101	5.2562	3.2623	8.4687
Total			1,540.9966	4,449.9665	718.5391

From the above:

$\Sigma(w_{A,t} \times w_{C,t}) = 1{,}540.9966$; $\Sigma w_{A,t}^2 = 4{,}449.9665$; and $\Sigma w_{C,t}^2 = 718.5391$

Hence:

$\sqrt{\Sigma w_{A,t}^2} = 66.7081$ and $\sqrt{\Sigma w_{C,t}^2} = 26.8056$

Therefore: cosine similarity for shared terms $= \dfrac{\Sigma(w_{A,t} \times w_{C,t})}{\left(\sqrt{\Sigma w_{A,t}^2} \times \sqrt{\Sigma w_{C,t}^2}\right)}$

$= 1{,}540.9966 \div (66.7081 \times 26.8056)$

$= 0.8618$

Source: Original.

Appendix B: Chapter 14

Computation of cosine similarity for all terms

Term (t)	$w_{A,t}$	$w_{C,t}$	$w_{A,t} \times w_{C,t}$	$w_{A,t}^2$	$w_{C,t}^2$
Energy	25.5150	8.4960	216.7755	651.0152	72.1821
Building	24.4449	0.0000	0.0000	597.5517	0.0000
CO2	23.3364	0.0000	0.0000	544.5890	0.0000
Heat	19.9416	0.0000	0.0000	397.6658	0.0000
Heating	18.9874	0.0000	0.0000	360.5195	0.0000
Reduction	18.9874	0.0000	0.0000	360.5195	0.0000
Renewable	18.9020	4.5845	86.6563	357.2845	21.0178
Emissions	17.5746	8.2349	144.7255	308.8648	67.8143
Carbon	17.0672	13.8532	236.4352	291.2903	191.9103
New	16.7210	2.8943	48.3958	279.5908	8.3771
Solar	16.7210	2.8943	48.3958	279.5908	8.3771
Technologies	16.5272	0.0000	0.0000	273.1490	0.0000
Cooling	16.1236	0.0000	0.0000	259.9705	0.0000
Low	15.4379	2.2923	35.3877	238.3300	5.2544
Savings	14.5310	0.0000	0.0000	211.1488	0.0000
Gas	14.5211	0.0000	0.0000	210.8635	0.0000
Measures	14.5211	0.0000	0.0000	210.8635	0.0000
Design	14.5112	0.0000	0.0000	210.5761	0.0000
Homes	14.5112	0.0000	0.0000	210.5761	0.0000
Electricity	14.4494	4.4891	64.8654	208.7863	20.1523
Buildings	13.8611	0.0000	0.0000	192.1298	0.0000
Efficiency	13.0860	0.0000	0.0000	171.2429	0.0000
Zero	12.8989	0.0000	0.0000	166.3811	0.0000
Levels	12.7791	0.0000	0.0000	163.3044	0.0000

(Continued)

Performance	12.7791	0.0000	0.0000	163.3044	0.0000
Sustainable	12.6433	1.4472	18.2968	159.8520	2.0943
Domestic	12.2366	0.0000	0.0000	149.7344	0.0000
Change	11.7680	5.0706	59.6708	138.4863	25.7109
Fuel	11.7402	6.0211	70.6883	137.8316	36.2532
Power	11.7402	2.9101	34.1652	137.8316	8.4687
Sector	11.7402	2.9101	34.1652	137.8316	8.4687
Water	11.7402	0.0000	0.0000	137.8316	0.0000
Significant	11.4391	0.0000	0.0000	130.8539	0.0000
Thermal	11.0752	0.0000	0.0000	122.6598	0.0000
Dwellings	11.0721	0.0000	0.0000	122.5914	0.0000
Space	10.5719	0.0000	0.0000	111.7651	0.0000
Policy	10.4803	3.8801	40.6653	109.8375	15.0555
Potential	10.4803	3.3804	35.4277	109.8375	11.4271
Dwelling	10.2803	0.0000	0.0000	105.6844	0.0000
Emission	10.2803	0.0000	0.0000	105.6844	0.0000
Generation	9.9340	0.0000	0.0000	98.6842	0.0000
House	9.9340	0.0000	0.0000	98.6842	0.0000
Wind	9.9340	0.0000	0.0000	98.6842	0.0000
Construction	9.6108	2.2923	22.0305	92.3678	5.2544
Air	9.2523	0.0000	0.0000	85.6043	0.0000
Housing	9.2523	0.0000	0.0000	85.6043	0.0000
Site	9.2523	0.0000	0.0000	85.6043	0.0000
Standard	9.2523	0.0000	0.0000	85.6043	0.0000
Ventilation	9.2523	0.0000	0.0000	85.6043	0.0000
Consumption	9.0309	2.9101	26.2809	81.5572	8.4687
Cycle	9.0309	0.0000	0.0000	81.5572	0.0000
Demand	9.0309	1.4472	13.0691	81.5572	2.0943
Embodied	9.0309	0.0000	0.0000	81.5572	0.0000
Model	8.8577	0.0000	0.0000	78.4585	0.0000
Government	8.6497	3.4384	29.7411	74.8180	11.8225
Technology	8.6497	1.4472	12.5175	74.8180	2.0943
Compared	8.2242	0.0000	0.0000	67.6380	0.0000
Build	8.1278	0.0000	0.0000	66.0613	0.0000
Study	8.1278	2.9101	23.6528	66.0613	8.4687
Residential	7.9744	0.0000	0.0000	63.5903	0.0000

Environmental	7.7505	4.5845	35.5321	60.0698	21.0178
Future	7.7505	0.0000	0.0000	60.0698	0.0000
Stock	7.7505	0.0000	0.0000	60.0698	0.0000
Wall	7.7505	0.0000	0.0000	60.0698	0.0000
Costs	7.6887	2.9101	22.3748	59.1154	8.4687
Climate	7.2247	6.7337	48.6490	52.1966	45.3426
Impact	7.2247	0.0000	0.0000	52.1966	0.0000
Increase	7.1962	2.2923	16.4955	51.7853	5.2544
Results	7.1962	0.0000	0.0000	51.7853	0.0000
Existing	6.7276	0.0000	0.0000	45.2603	0.0000
High	6.7276	1.4472	9.7359	45.2603	2.0943
Household	6.6453	0.0000	0.0000	44.1599	0.0000
Hot	6.6433	0.0000	0.0000	44.1329	0.0000
Period	6.6433	0.0000	0.0000	44.1329	0.0000
Research	6.6433	2.2923	15.2281	44.1329	5.2544
Higher	6.1682	0.0000	0.0000	38.0463	0.0000
System	6.1682	6.7608	41.7017	38.0463	45.7082
Current	6.0206	0.0000	0.0000	36.2476	0.0000
Built	5.5361	0.0000	0.0000	30.6478	0.0000
Control	5.5361	0.0000	0.0000	30.6478	0.0000
Internal	5.3162	0.0000	0.0000	28.2624	0.0000
Required	5.3162	0.0000	0.0000	28.2624	0.0000
Available	4.8165	1.4472	6.9702	23.1985	2.0943
Data	4.8165	1.4472	6.9702	23.1985	2.0943
Effect	4.8165	0.0000	0.0000	23.1985	0.0000
Level	4.8165	3.4384	16.5609	23.1985	11.8225
SAP	4.8165	0.0000	0.0000	23.1985	0.0000
Assessment	3.0103	0.0000	0.0000	9.0619	0.0000
National	3.0103	4.4891	13.5136	9.0619	20.1523
Scenario	3.0103	0.0000	0.0000	9.0619	0.0000
Scenarios	3.0103	0.0000	0.0000	9.0619	0.0000
Annual	1.8062	0.0000	0.0000	3.2623	0.0000
Average	1.8062	2.9101	5.2562	3.2623	8.4687
London	1.8062	0.0000	0.0000	3.2623	0.0000
Models	1.8062	0.0000	0.0000	3.2623	0.0000
Office	1.8062	0.0000	0.0000	3.2623	0.0000

(Continued)

Temperatures	1.8062	0.0000	0.0000	3.2623	0.0000
Weather	1.8062	0.0000	0.0000	3.2623	0.0000
Project	0.0000	20.2011	0.0000	0.0000	408.0836
Uganda	0.0000	14.7182	0.0000	0.0000	216.6239
CDM	0.0000	11.3731	0.0000	0.0000	129.3479
Forest	0.0000	7.3591	0.0000	0.0000	54.1560
Development	0.0000	7.2247	0.0000	0.0000	52.1966
Improved	0.0000	6.8768	0.0000	0.0000	47.2899
Diesel	0.0000	6.6227	0.0000	0.0000	43.8596
Land	0.0000	6.0211	0.0000	0.0000	36.2532
Rural	0.0000	6.0211	0.0000	0.0000	36.2532
Biomass	0.0000	5.9855	0.0000	0.0000	35.8263
Market	0.0000	5.8202	0.0000	0.0000	33.8750
Stoves	0.0000	5.8202	0.0000	0.0000	33.8750
Wood	0.0000	4.8502	0.0000	0.0000	23.5243
Charcoal	0.0000	4.5845	0.0000	0.0000	21.0178
Briquettes	0.0000	4.3415	0.0000	0.0000	18.8484
International	0.0000	4.3415	0.0000	0.0000	18.8484
Local	0.0000	4.3415	0.0000	0.0000	18.8484
Global	0.0000	4.2255	0.0000	0.0000	17.8548
Many	0.0000	4.2255	0.0000	0.0000	17.8548
Small	0.0000	4.2255	0.0000	0.0000	17.8548
Countries	0.0000	3.8801	0.0000	0.0000	15.0555
Activities	0.0000	3.7409	0.0000	0.0000	13.9946
Benefits	0.0000	3.4384	0.0000	0.0000	11.8225
Private	0.0000	3.4384	0.0000	0.0000	11.8225
Support	0.0000	3.3804	0.0000	0.0000	11.4271
Capacity	0.0000	2.9101	0.0000	0.0000	8.4687
Generator	0.0000	2.9101	0.0000	0.0000	8.4687
Grid	0.0000	2.9101	0.0000	0.0000	8.4687
Implementation	0.0000	2.9101	0.0000	0.0000	8.4687
Supply	0.0000	2.9101	0.0000	0.0000	8.4687
Cooking	0.0000	2.8943	0.0000	0.0000	8.3771
Country	0.0000	2.8943	0.0000	0.0000	8.3771
Health	0.0000	2.8943	0.0000	0.0000	8.3771
Africa	0.0000	2.2923	0.0000	0.0000	5.2544
Community	0.0000	2.2923	0.0000	0.0000	5.2544

Fuelwood	0.0000	2.2923	0.0000	0.0000	5.2544
Human	0.0000	2.2923	0.0000	0.0000	5.2544
Impacts	0.0000	2.2923	0.0000	0.0000	5.2544
Production	0.0000	2.2923	0.0000	0.0000	5.2544
City	0.0000	1.4472	0.0000	0.0000	2.0943
Economic	0.0000	1.4472	0.0000	0.0000	2.0943
Kampala	0.0000	1.4472	0.0000	0.0000	2.0943
Needs	0.0000	1.4472	0.0000	0.0000	2.0943
Plan	0.0000	1.4472	0.0000	0.0000	2.0943
Planning	0.0000	1.4472	0.0000	0.0000	2.0943
Resources	0.0000	1.4472	0.0000	0.0000	2.0943
Social	0.0000	1.4472	0.0000	0.0000	2.0943
Technical	0.0000	1.4472	0.0000	0.0000	2.0943
Total			1,540.9966	12,051.2111	2,182.8028

From the above:

$$\Sigma(w_{A,t} \times w_{C,t}) = 1{,}540.9966;\ \Sigma w_{A,t}^2 = 12{,}051.2111;\ \text{and}\ \Sigma w_{C,t}^2 = 2{,}182.8028$$

Hence:

$$\sqrt{\Sigma w_{A,t}^2} = 109.7780 \text{ and } \sqrt{\Sigma w_{C,t}^2} = 46.7205$$

Therefore: cosine similarity for all terms $= \dfrac{\Sigma(w_{A,t} \times w_{C,t})}{\left(\sqrt{\Sigma w_{A,t}^2} \times \sqrt{\Sigma w_{C,t}^2}\right)}$

$$= 1{,}540.9966 \div (109.7780 \times 46.7205)$$
$$= 0.3005$$

Source: Original.

Index

Note: **Bold** page numbers refer to tables and *italic* page numbers refer to figures.